S0-BZL-777

Organic form

Organic form

The life of an idea

edited by

G. S. Rousseau

Associate Professor of English,
University of California, Los Angeles

Routledge & Kegan Paul

London and Boston

First published 1972
by Routledge & Kegan Paul Ltd
Broadway House, 68-74 Carter Lane,
London EC4V 5EL
and
9 Park Street,
Boston, Mass. 02108, U.S.A.
Printed in Great Britain by
C. Tinling & Co. Ltd
London and Prescot

ISBN 0 7100 7246 5

Contents

Figures

Acknowledgments

The editor and authors would like to thank Frances Funaro and Roger Hambridge, both of the University of California at Los Angeles, for various types of research assistance and editorial expertise in preparing the manuscript for publication; and Ellen Cole and Betty Baker for typing and proof-reading the entire manuscript.

Notes on the contributors

G. N. GIORDANO ORSINI (b. 1903), Professor of Comparative Literature at the University of Wisconsin at Madison, is a world famous aesthetician and literary historian who has made the study of Coleridge and organic form his particular province of expertise. Trained in Italy by Benedetto Croce and at the University of Florence, Orsini came to the United States in 1949, where he has since lived. For many years he has been at work on a definitive history of organic form, portions of which have appeared in the journal *Comparative Literature*, and it is fair to say that he probably knows more about the genesis of this notion than any man alive. Professor Orsini's history, when published, will give greater context to the present book and systematically demonstrate how the idea progressed and grew, and how it spread from country to country, thinker to thinker.

PHILIP C. RITTERBUSH (b. 1936) is Chairman of Organization:: Response and was formerly Director of Academic Programs at the Smithsonian Institution. He is a professional historian of biology. Trained at Yale and Oxford universities in history and history of science, Dr Ritterbush is the author of several works dealing with organic form including *Overtures to Biology: The Speculations of Eighteenth-Century Naturalists* (Yale University Press, 1964), *The Art of Organic Forms* (Smithsonian Institution Press, Washington, 1968), 'The Shape of Things Seen: The Interpretation of Form in Biology' in *Leonardo III* (1970). He has been especially interested in organic form as it appears in theories of biology and painting, as his essay shows. At present Dr Ritterbush is Chairman of Organization:: Response a Washington-based research organization with consultants throughout the United States which explores the cultural role of institutions of learning.

G. S. ROUSSEAU (b. 1941), Associate Professor of English and Fellow of the Humanities Institute at the University of California at Los Angeles, is a disciple of Marjorie Hope Nicolson and, more generally, of the Lovejoy school of the history of ideas. He has devoted himself to studying the interrelations of literature and science, particularly in the last four centuries. Among his publications on this subject are *This Long Disease, My Life: Alexander Pope and the Sciences* (Princeton University Press, 1968) and *Bishop Berkeley and Tar Water* (Oxford University Press, 1970), both written in collaboration with Marjorie Hope

Nicolson; and 'Science and the Discovery of the Imagination in Enlightened England', ECS (University of California, 1969). Professor Rousseau is at present chairman of the Literature and Science Section of the Modern Language Association of America. He is working on a three-volume history of the idea 'imagination', in which, he believes, the notion of organic form plays a large role.

WILLIAM K. WIMSATT (b. 1907), Professor of English at Yale University, is one of America's leading literary critics. The author or editor of a dozen books dealing with problems in criticism and interpretation, Wimsatt is especially known for his collaborative effort written with Cleanth Brooks entitled *Literary Criticism: A Short History* (Alfred A. Knopf and Routledge & Kegan Paul, 1957), and for *The Verbal Icon* (University of Kentucky Press, 1954; Methuen, 1970), and *Hateful Contraries* (University of Kentucky Press, 1965). He has also written articles and books on eighteenth-century English authors, particularly Alexander Pope, Samuel Johnson, and James Boswell. His essay in this volume demonstrates once again his piercing quality of mind and powerful logical faculty, as have earlier essays on the intentional fallacy, the affective fallacy, and the domain of criticism.

G. S. Rousseau

Introduction

The essays in this volume were originally delivered in December 1970 at the Literature and Science section of the annual Modern Language Association meeting in New York City. They appear here for the first time in print in somewhat expanded form. To them has been appended a fairly representative, although by no means exhaustive, bibliography of secondary sources dealing with the aesthetic problem of organic form. Since a description of the original meeting in New York offers a context into which these papers may be read and understood, a brief word on that subject follows.

The Literature and Science section of MLA, ever since its early days in the 1960s when Marjorie Hope Nicolson and her students were active at its gatherings, has sought to focus each meeting on a particular topic (e.g. the idea of the Great Chain of Being, Romanticism and the natural sciences, time and energy as literary and scientific concepts). When its Executive Board selected me in 1968 to be the Chairman for the 1970 meeting, I promised myself to discover a genuinely important topic among the interrelations of literature and science that had not previously been treated. Little did I know at that time (in the autumn of 1968) that fortune would favor me with a panel of three of the best thinkers of these times. When I finally decided upon organic form, after having thrown out of court several dozen possibilities, I then set out to locate three scholars – the fixed number of panelists who may appear at a section of MLA. Experience at previous meetings convinced me that one of the three must be a practicing scientist or historian of science: how many gatherings had failed because the literary critics spouted jargon at each other without the slightest attempt to be understood by the 'other culture'. But I also felt that a stimulating session required the acumen of a highly discriminating literary theorist if not philosopher of literary form. Finally, looking over the horizon as Hannibal over the Alps, I hoped to discover the man most enlightened in the subject of organic form, its genesis, history, and reception in modern times, not only on this continent but throughout the world. When I say that fortune was indeed benevolent, readers will I hope understand that each of my three searches was fulfilled.

Ritterbush, Wimsatt and Orsini formed a team upon which no chairman could hope to improve: the first, an historian of science who had written two books on the matter; the second, one of America's leaders in the field of literary criticism (who had in fact been one of Ritterbush's teachers at Yale); and the third, Professor Orsini, that scholar who answered to my search for peerless competence in the history and life of this idea. These men were (naturally) known to each other, surely by reputation if not personally, as was confirmed by the difficulty I encountered during the months we were planning the program, or imposing an edict of non-communication with each other about their papers. And yet I was determined that each should think and write independently without seeing the other papers. This edict of non-communication was crucial for many reasons including spontaneous give and take at the meeting – especially in the case of disagreement – but also to guarantee total independence of mind; for I feared that all three papers, however excellent, would say the same thing, would be monotonously repetitive and redundant. The outcome did not, however, bear out my apprehension.

Professor Orsini, asked to appear first because he had elected to focus on 'The Ancient Roots of a Modern Idea', discusses in this expanded version of his talk the 'birth' of the idea in ancient Greece. By centering his attention on the Platonic and Aristotelian phase of the idea of organic form, he shows not only its rise but the mould into which this aesthetic precept was cast by two philosophers who have had monumental influence on aesthetics and literary criticism of the last twenty centuries. Trained in the school of Benedetto Croce and the author of a volume about his work (*Benedetto Croce: Philosopher of Art and Literary Criticism*, 1961), Professor Orsini is understandably impressed by the idea of organic form as a tenable aesthetic precept as well as a metaphor used to describe the parts and whole of a work of art. His inclination – as his essay shows – not only has the support of Croce but of Coleridge before him, to say nothing of the mighty Greeks.

Dr Ritterbush, the second speaker, demonstrated an altogether different perspective. An historian of biology who has been immersed for many years in the aesthetic relations between biology and painting, he focused on the Romantic period, especially on Goethe in Germany and Coleridge in England, in an attempt to show how these men built on older notions but also supplemented them by their empirical conviction that in both biology and art the sum of the parts does not equal the whole. Ritterbush then abundantly demonstrates how this idea (not merely an aesthetic bias or preference, but a powerful idea applicable in many areas besides art) took root among nineteenth-century scientists, engineers, philosophers, and critics. He cites the well-known remark of T. H. Huxley which epitomizes the notion for the Victorians as well as for the latter nineteenth century:

> In travelling from one end to the other of the scale of life, we are taught one lesson, that living nature is not a mechanism but a poem; not merely a rough engine-house for the due keeping of pleasure and pain machines, but a palace whose foundations, indeed, are laid on the strictest and safest mechanical principles, but whose superstructure is a manifestation of the highest and noblest art.[1]

Ritterbush, whether discussing individual cells, the mind of man, or a great painting by Dürer, shows how the whole is always more than the sum of its parts. If he has laid considerable emphasis on Goethe in demonstrating this principle, it is because the German poet and scientist is essential to the vitality of the idea at its most critical moment and because Goethe expresses so well the strengths and weaknesses of the argument. It is fair to say that Ritterbush's contribution is among the most eloquent statements uttered by a modern historian of science whose aim is to trace objectively and dispassionately the scientific dimensions of the concept of organic form.

Professor Wimsatt's paper took the chairman and other panelists by surprise. Writing from his *ne plus ultra* vantage as America's leading literary theorist, he was more critical of some exponents of the idea – a sophistic metaphor as he calls it. Abandoning for the present historical perspective and ranging all over the vast expanse of European literature and philosophy, he squarely and boldly points out a logical difficulty hitherto unnoticed or ignored. 'We have been skirting a sophism,' he writes, 'namely, the notion that the representation of biological forms in a work of verbal or visual art implies something about the presence of organic or artistic form in that work.' With characteristic ingenuity Wimsatt builds on this differ-

[1] 'On Natural History, as Knowledge, Discipline and Power, 1856', in *The Scientific Memoirs of Thomas Henry Huxley*, ed. M. Foster and E. R. Lankester, London: Macmillan, 1898–1902, Vol. I, p. 311.

ence he has noticed in each realm and unfolds some of the most penetrating and provocative questions ever asked about organic form. He challenges not only Romantic analogies between plant life and the life of the mind and between plant life and works of art, but also defies Plato and Aristotle. 'Plato said a composition should have an organized sequence of parts, and that it ought to be like a living being, with foot, body, and head. And Aristotle said that it ought to be a unity, like an organism. But we might ask of Plato, what are the foot, body, and head of a poem? Or of Aristotle, what are the beginning, middle, or end of a squirrel or tree?' Like Orsini, Wimsatt agrees that the simile can be carried too far: there is little point, as both scholars note, in asking what corresponds to the stomach in a tragedy. The main thrust of Wimsatt's essay, therefore, is determination of a sensible middle ground, neither searching for the liver of a comedy nor discarding the metaphor altogether. His conclusion is as balanced and prudent as the rest of the essay and, if followed, would probably enhance the likelihood of superior criticism. Avoiding the shoals of either extreme he advocates 'that homelier and humbler sort of organicism, in the middle, which I have been trying to describe – empirical, tentative, analytic, psychological, grammatical, lexicographic'.

Finally, a word on the bibliography. As I have cautioned the reader in the explanatory note, it is hardly exhaustive but it does demonstrate abundantly how much ink has been spilled on this vexed question. No doubt this book will have successors, yet I am skeptical that their contents will be more varied in approach and in the diversity of their conclusions than in the following essays. Perhaps this diversity of opinion prognosticates some radical change in the direction of modern literary criticism – for we are, as Wimsatt indicates, at a juncture in critical history. As this manuscript was completed early in 1971 later books have not been consulted.

G. N. Orsini

The ancient roots of a modern idea

I propose to speak of a principle that in 1926 an American aesthetician, Professor De Witt Parker, called 'the ancient principle of organic unity', and designated as 'the master principle of aesthetic form, which all the other principles subserve'.[1] More recently, in 1967, the article on 'Problems of aesthetics' by Professor John Hospers in the *Encyclopaedia of Philosophy* (I, 43) asked: 'What, then, are the principles of art by which a work of art is to be judged, at least in its formal aspect?' and answered: 'the central criterion, the one most universally recognized, is unity (sometimes called organic unity).' Hospers formulated the criterion thus: 'The unified object should contain within itself a large number of diverse elements, each of which contributes to the total integration of the unified whole, so that there is no confusion despite the disparate elements within the object. In the unified object, everything that is necessary is there, and nothing that is not necessary is there.' In another work, Hospers enunciated an important corollary to this principle: 'no part could be removed without damaging the remaining parts.'[2]

For instance, all the lines, and indeed all the words, of a good poem are necessary to it, and none can be removed without altering the whole. The same applies to a play – all of its scenes are necessary to the dramatic effect, and none can be removed without altering the whole for the worse. All the episodes of a great tragedy are necessary to it, e.g. in *Hamlet*, the scene with the players, the baiting of Rosencrantz and Guildenstern, the churchyard scene, etc. The question can be raised about any great poem – does the *Divine Comedy* possess organic unity? Does Goethe's *Faust*, composed of two diverse parts composed after an interval of time, possess it? Different answers have been given to these questions and the debate still goes on. But the principle of organic unity implicit in these questions deserves critical and historical attention.

Works of art sometimes contain parts which do not seem indispensable. So it may be that organic unity is not so much an essential property of every work of art, as a counsel of perfection, a perfect harmony that we cannot expect every work to achieve.[3]

On the other hand, the history of the principle shows how widespread and enduring it is. Erwin Panofsky applied it also to art criticism, and assigned an important role to it, 'that great principle of classical and Renaissance aesthetics', i.e. that

> beauty is the harmony of the parts in relation to each other and to the whole. This concept, developed by the Stoics, unquestioningly accepted by a host of followers from Vistruvius and Cicero to Lucian and Galen, surviving in medieval scholasticism and ultimately established as an axiom by Alberti, who does not hesitate to name it the 'absolute and primary law of nature', was the principle called *symmetria* or *harmonia* in Greek, *symmetria*, *concinnitas* and *consensus partium* in Latin, *convenienza*, *concordanze* or *conformità* in Italian. . . It meant, to quote Lucian, 'equality or harmony of all parts in relation to the whole'.[4]

The historical development of this principle will now be taken up in connection with its almost inseparable opposite, i.e. 'Mechanical unity'. As S. H. Butcher, the commentator of Aristotle's *Poetics*, says: 'Organic as distinct from mechanical unity: not the homogeneous sameness of the sandheap, but unity combined with variety.'[5] We might say that a mechanical unity exists when the parts of a work are just placed one next to the other and only extrinsically joined, being made to enter in a pre-established framework without any intrinsic connection. We are all acquainted with loosely-jointed works of literature, and with works contrived in order to conform to some pre-established scheme, regardless of the nature of the subject or of its elements. Now the roots of this distinction between the organic and the mechanical are to be found in the history of philosophy, and were outlined by Adolf Trendelenburg in the last century.[6] Following him, Rudolf Eucken traced the parallel history of the ideas and the terms in a useful survey which I shall adopt, while making some additions to it.[7]

> The terms organic and mechanical, like the ideas they stand for, are old, but it was long before the terms became associated with the ideas. 'Mechanical' is a well-established expression in Aristotle for the designation of the art of invention, i.e. the construction of machines (*hè mechaniké, tà mechaniká*). The word continued to bear this meaning throughout the centuries, and since the time of Descartes it was served to denote a theory which explains the operations of nature as the result of the combination of minute particles of matter originally endowed

> with motion. So the works of nature appear to differ from those of man solely in the greater complexity of their structure, which is to say quantitatively and not qualitatively.

I will here interrupt to point out the analogy with literature. In a literary composition, the minute particles may be identified as the single words of which the composition is made up. For, according to the common view, each word is endowed with its own power of signification and conveys its own idea. Or, taking another view, the elementary particles of which a work is composed could be the images which are to be found in it, each again endowed with its own particular power of expression. In a mechanic unity, these components are merely put together, not organically united.

The analogy between mechanism and literary composition was drawn later, as Eucken proceeds to say:

> the transference to mental phenomena was not at first thought of, and 'mechanical' was frequently reckoned as synonymous with 'material.' Finally the term 'mechanical' was applied to psychological phenomena, at first figuratively and then literally. At the end of the 18th century, Kant in his philosophy of nature clearly worked out the contrast between mechanical and dynamical explanation.
>
> The term 'organic' was at first used by Aristotle, but not in the modern sense. Corresponding to the word *órganon*, instrument, organic meant 'instrumental'. It was used of the living body as a whole, but more frequently of the members of the body, which are as it were its tool.
>
> The term retained its meaning without change through the Middle Ages and in the modern world until well into the 18th century.

I will here interject an example from Milton: the 'serpent's tongue' is said to be 'organic' (P.L., IX, 530) because it is used as a tool.

> Then, in the 18th century, came the classical period of German thought, and with it the impulse to endow nature with a life of its own. This trend first added the property of life to the term 'organic' and made it its main characteristic. Kant, with his exact concepts and distinctions, exerted a special influence in

> this direction, though Herder, Jacoby and others should not be forgotten. Indeed, Kant formulated the classic teleological definition of the living organism: 'one in which every part is reciprocally means and end'

—a definition later echoed by Coleridge.[8]

> The new meaning of organism was then transferred from natural beings to society and to the State, then to law, history, and so forth. 'Organic' became a favorite term of the Romantic school, though at the same time we find it spreading beyond specific schools and trends of thought, and passing into ordinary speech. Thus, while 'mechanic' and 'organic' originally meant almost the same thing, they came finally to stand in almost complete opposition to each other.
>
> For the two terms serve to designate a contrast in the nature of things which has long been a problem. The artistic and creative way of thought characteristic of the classic age of Greece ranked the whole above the parts and the living above the lifeless, explaining the latter through the former. It was in sympathy with this whole trend that the idea of the organism, though not the term, was adopted by Aristotle. He also originated the formula that in an organic being the whole precedes the parts.

For Aristotle, too, a living body exhibits teleology more definitely than an artifact.

Now in this survey of Eucken's one name is missing and perhaps the most important one, for it is the name of the originator of the principle of organic unity in art, namely Plato. For it was Plato who first enunciated it in his dialogue *Phaedrus*. It is in this dialogue that Socrates says that a composition 'should be like a living being, with a body of its own as it were, and neither headless nor footless, but with a middle and members adapted to each other and to the whole' (246 C, Jowett's translation slightly modified). As Butcher, the commentator of Aristotle's *Poetics*, said in 1904: 'Here, for the first time, the law of internal unity is enunciated as the primary condition of literary art, now a commonplace, then a discovery' (loc. cit.). It is also worth noting that there is a double relationship required here,

the relation of the parts to each other and their relation to the whole. And they must be adapted or in keeping with each other.

I will do now what I have not yet done in any of my previous papers on this subject.[9] I will place Plato's formulation in its context and relate the whole discussion in which it occurred and in which Plato sets up a standard of criticism by which compositions are to be judged. He also provides examples of compositions, both in prose and verse, that fail to meet that standard. True, the principal example given here is in prose, but Plato immediately extends it to verse, and there is also a clear indication that it applies to dramatic composition too. Later in the dialogue, contemporary handbooks of rhetoric are surveyed and criticized; in effect they reject organic unity. So, in addition to all the other philosophical theories and fancies that the *Phaedrus* contains, it may be said to present also a primer of literary criticism, the earliest that we have in Greek literature.

At this point someone may ask, if this is so, how is it that this dialogue does not occupy a place in the history of criticism comparable to that of Aristotle's *Poetics*, or of Horace's *Ars Poetica?* It seems to be barely mentioned in the general histories of aesthetics[10] and of literary criticism, while according to the above it should have an honorable place in them. Well, recognition of this dialogue is growing,[11] and there are some extrinsic reasons why it has not hitherto received the full attention it deserves. The main reason, perhaps, is that the *Phaedrus* has long been thought to refer *only* to rhetoric or the art of speech, and not to the higher arts of literature. Certainly what Socrates has to say applies to prose, and the main composition offered as an example (though not of good writing) is of the kind usually labelled an 'oration' (*lógos*, or speech), and its presumed author is Lysias, who is traditionally classified as one of the Attic orators. Actually the piece in question is not a spoken address, but a fully written-out prose composition, such as we would call an essay today. Lysias, as is recalled in this very dialogue (275 C), had a *penchant* for written composition and nourished literary ambitions. The essay consists of the defence of a paradox, like the defence of unworthy things cultivated as a fad by ancient writers and again in the Renaissance. The paradox of Lysias in this piece happens to be that a suitor who is not in love is preferable to

a suitor who is in love, and various ingenious arguments are brought forward in support of this unusual preference.

Now Socrates' disciple Phaedrus is much impressed by this essay and ends by reading it all out to Socrates, so that we get it in its entirety. It resembles the known style of Lysias so closely that some critics consider it a genuine composition of Lysias, not an invention of Plato's. However that may be, it certainly provides good material for the exercise in practical criticism which Socrates then performs upon it. His analysis begins by finding fault with its manner: it is repetitious, saying the same things over and over, with different words but no difference in meaning (235 A). Phaedrus replies that this is just its merit: it says *all* that can possibly be said upon the subject, and even Socrates cannot find anything more to add to it. Socrates admits this fact, but observes that it is not possible to add to arguments that arise inevitably from the subject, and which do not call for any special invention. It has been noted that Socrates is following here the traditional distinction in Greek criticism between matter and arrangement, a distinction which will later become that of content and form.[12]

Socrates then proceeds to another kind of criticism, positive and not negative. To show Phaedrus how the thing should have been done, he goes on to rewrite the whole piece. It should have begun with a proper definition of its subject, love, and from this definition the reasons for rejecting the sincere lover should have been drawn (237 B – 241 D). But when Socrates reaches the second part of the argument, which consists of praising the non-lover, Socrates suddenly comes to a stop and gives up the whole discussion. He says that his demon – that kind of guardian angel that protected Socrates and warned him against ill-doing – has checked him, and he will not go on blaspheming true love. Instead, he sets upon an entirely new discourse, which will treat of love as the subject deserves (244 ff.). Sophistical ingenuity is now left behind; Socrates speaks out of deep conviction, and delivers one of his most eloquent and imaginative discourses, which climaxes in the description of the 'place above the heavens', where immortal souls reside before being born and enjoy the sight of the eternal Ideas. Love is now seen to be a kind of divine madness, akin to poetry and to religious inspiration (250).

Plato states definitely here that only the man who is possessed by a divine madness or inspiration can write poetry. For among the eternal Ideas visible in the hyperuranium there is also the idea of Beauty, which has the special property of being apparent even in material objects (250 D). It is the recollection of what we saw in heaven that makes us recognize beauty here on earth (250 A). Having shown how the proper praise of love should be conducted, Socrates turns to consider the art of writing. He formally asks Phaedrus, 'What, then, is the way of writing well or badly?' To make his intention quite clear, he adds, 'whether in a public or private document, in verse as a poet, or in prose as an ordinary man?' (258 A). So the principle is not to be restricted to oratory or even to prose alone, but to be extended to poetry, in its widest sense. The answer to the question of how to write well is the precept of organic unity which I have already quoted (264 C), although other qualities are also required (270 A and 274 B).

It is important to note the difference between this theory and the theory of the Idea of Beauty, expounded just before. According to the latter, a thing is beautiful only when the Idea of Beauty descends from above into it and becomes part of it – the so-called presence of the Idea in the particular. For the organic theory, a work acquires a quality, which is *not* designated explicitly as beauty but is akin to it, only when it is properly organized. This organization, far from being a passive reception of an Archetype from above, is a product of the unaided activity of the human mind. This is also different from the Inspiration theory presented in 245 A. It is not an insufflation from above or from beyond, but a result of the unaided activity of the mind, and therefore in this sense subjective and not objective. To confirm this, let us look again closely at the passage in which Plato presents this concept.

> Every discourse (*lógos*) must be composed like (*hósper*, in the likeness of) a living being, with a body of its own, as it were, so as not be headless or footless, but to have members and a middle (*ákra* and *mésa*) arranged in a fitting (*prépon*) relation to each other and to the whole (264 C).

Then a short poem is given as an example. This poem is an inscription upon a statue placed near a fountain on the grave of the legendary Midas. This little 'epigram', as it is called technically, is not without charm in the original. It may be that its Greek character has for us the attraction of the antique, but to Socrates it appears to have been just an inscription by the wayside, somewhat casually put together. Here it is in the translation by Hackforth, which has the merit of being, like the original, in verse:

> A maid of bronze I stand on Midas' tomb,
> So long as waters flow and trees grow tall,
> Abiding here on his lamented grave,
> I tell the traveller Midas here is laid.

Socrates points out that, the way this is written, the lines can be read in any order without making any difference to the sense: 1, 2, 3, 4 as well as 4, 3, 2, 1, or any other arrangement. The same is true, says Socrates, of Lysias' prose essay: he 'does not begin at the beginning', 'the parts of the essay are thrown out helter-skelter', 'the second topic is not placed second for any good reason, nor any of the others'.

Now this suggests another relationship between the parts of a composition which has not been mentioned here – the parts should be presented in their proper order. Order, or *táxis*, becomes prominent in other dialogues of Plato, and perhaps we may consider it implicit in this statement of doctrine.

Returning to the epitaph of Midas, Socrates' example of faulty composition, I cannot quite find it in me to consider it faulty, even if it is re-arrangeable at will. Being a short composition, each of its lines is complete in itself and does not cause confusion in re-arrangement. This introduces the element of size, which is a point of Aristotle's, made in his definition, 'Beauty is a matter of order and size' (*Poet.*, 1450 B 36). Furthermore, there is such a thing as progression through parallel statements. Each of the lines in the epitaph is complete in itself, and they all contribute something to the whole, which would be altered if one of them were omitted. The converse would be: is the poem altered by any addition? We need not attempt experiments and invent new verses, for an ancient author has accommodatingly provided us with a version of this same epitaph that has two more lines (Diogenes Laertius, I, ch. 90), which add nothing to it and appear mere padding. Hence the epigram would seem to satisfy the conditions laid down by Plato for unity: each of the parts is in keeping with the others and with the whole, so that any addition or diminution would spoil it. In prose, particularly argumentative prose like Lysias' essay, such looseness of structure is not to be tolerated. In poetry, which is not logical argument, such looseness is not necessarily bad, because much depends upon other factors.

The parallel progression which we found in the epigram does not involve complete lack of form, and therefore does not support some modern theories about so-called 'open form'. But Socrates' criticism of the essay by Lysias is fully justified, for the piece is truly disorganized. In Socrates' definition, the relation of the parts to each other and to the whole is termed 'fitting, proper' (*prépon*). This is a general term which can be particularized in various different ways. It is not particularized here, so it cannot be charged with being tautological: Socrates is not saying that 'beauty consists of a beautiful arrangement', but defines only the structure of the arrangement, leaving its quality open. For that arrangement can be defined as involving suitability of means to end, which could make it consistent with an utilitarian aesthetic ('beautiful is what satisfies a purpose'), such as is placed in the mouth of Socrates in other dialogues.

However we conceive it, there is still an essential part of the definition of organic unity which is lacking in this statement – 'any alteration of the parts involves an alteration of the whole.' That addition we know is by Aristotle (*Poet.*, 1451 A 32) and it has now become an integral part of the definition, as we saw in the quotations from modern critics made at the beginning.

Now Plato has made it clear that this principle applies not only to prose but also to poetry. He refers to it explicitly, also to dramatic composition.

'Suppose', says Socrates, 'a person were to come to Sophocles or to Euripides and to say he knows how to make a very long speech (even) about a small matter, and a short speech (even) about a great matter, and also a sorrowful speech, or any kind of speech, and in teaching this fancies that he is teaching the art of tragedy?'

Phaedrus, who by this time has learnt his lesson, replies: 'They too would surely laugh at him if he fancies that tragedy is anything but the arranging of those elements in such a manner that will be suitable to each other and to the whole' (268).

So tragedy is also subject to the law of organic unity. Its parts, as here defined, are the speeches of which it is made up. Surely a rather vague and superficial definition of the parts of tragedy. But within this narrow frame of reference, Plato applies organicity to tragedy more definitely than Aristotle does in his more specific division of tragedy into parts in the *Poetics*. For Aristotle saw organicity only in one of these parts, i.e. the plot or *mythos*, and did not apply it to the unification of all the parts into a single composition, but limited himself to enumerating them, while stressing the pre-eminence of plot.

Plato's organicism is also the foundation for his critique of the systems of rhetoric which were already flourishing in his time, with all their divisions and subdivisions. The devices of rhetoric were defined with Greek subtlety by a crowd of writers. The art of public speaking was much in demand and paid well, so handbooks were prepared teaching its technique. This technique is criticized by Socrates in the *Phaedrus* on the general ground that they set pre-established formulas and prescriptions for composition, regardless of the subject, the capacity of the author, and the audience. The survey is spiced with irony throughout:

> PHAEDRUS There is surely a great deal to be found in the books on rhetoric.
> SOCRATES Yes, thank you for reminding me. The first point, I suppose, is that a speech should begin with a Preamble: that is what you mean, the niceties of art?
> PH. Yes.
> SOC. And next comes Exposition accompanied by Direct Evidence, thirdly Indirect Evidence, fourthly Probabilities; the great Byzantian word-maker also speaks of Confirmation and Supplementary Confirmation.
> PH. You mean the excellent Theodorus?
> SOC. Of course. And we are to have a Refutation and a Supplementary Refutation, both for the prosecution and for the defence. And can we leave the admirable Evenos of Parus out of the picture, the inventor of Covert Allusion and Indirect

> Censure and (according to some accounts) the inventor of Indirect Censure in Mnemonic Verse? A real master, that . . . And then Pulus: what are we to say of his *Muses' Treasury of Phrases* with his Reduplications and Maxims and Similes, and of words *à la* Lycumnius which that master made him a present of as his contribution to fine writing?
> PH. But didn't Protagoras in point of fact produce many such works, Socrates?
> SOC. Yes, my young friend: there is his *Correct Diction* and other excellent works. On the way to conclude a speech there seems to be general agreement, though some call it Recapitulation and some other name.[13]

It should be apparent by now that what Socrates is here attacking is the system of Rules – the Rules for Fine Writing. 'Hence,' says Butcher, the Aristotelian commentator, 'the uselessness of mere mechanical rules. All the "ologies" and technical terms of the rhetoricians will not teach you how to speak or to write well' (op. cit.). Then follows the positive teaching of Plato on this topic, which may be summarized thus: the subject of a composition should determine its divisions and its arrangement; the subject, we may say, as conceived by the mind of the author, in its particularity, not as an abstract scheme. The rhetorician may at best be said to deal with the preliminaries of composition rather than composition itself, Plato concludes (268 E, 269 A and C).

Again in the *Statesman* (299 D-E) Plato argues against general rules set up for *all* the arts and crafts, specifically including 'all the imitative arts, like painting' (299 D).

Plato therefore, I make bold to say, by rejecting the Rules anticipates the criticism of the Rules of Poetry made in the eighteenth century by those critics who are now termed Pre-Romantic. Romantic critics like Schlegel rejected the Rules on the same ground as Plato – they violated organic unity.[14]

To extend the application of Plato's critique of rhetoric to poetics, we might perhaps indulge the imagination a little and make a modernized Socrates speak like this to his disciple:

> When it comes to the dissection of a work of literature, let us not forget the mighty man from Stagira, with his six parts

of tragedy, three kinds of Peripeties, and five kinds of Recognition. Nor shall we omit the poet Horace, who set up a lot of rules for writing plays, an art which he never cultivated himself; but he required three actors and no more, five acts and no less, and would have only traditional subjects. Let us also cast a look at the mass of dramatic rulemongers and pigeonholers that followed them through the ages, multiplying classifications but not adding to the number of good plays. A special niche should be found, I profess, for the French nineteenth-century critic, Nepomucème Lemercier, with his enumeration of twenty-six rules for Tragedy, twenty-three rules for Comedy, and twenty-four for the epic poem.[15] A real master was also Gustav Freitag, with his seven parts of tragedy, improving upon Aristotle's six (which he later reduced to four). Freitag's are famous, i.e. Exposition, Initial Impulse, Rising Action, Crisis, Falling Action, Moment of Final Suspense, and Catastrophe. Can we leave our contemporaries out of the picture, like the subtle Empson with his *Seven Types of Ambiguity*, and the prolific Northrop Frye with his seven classes of imagery, four 'pre-generic narrative elements of literature', 'six phases of tragedy', and so forth? or Wayne Booth, author of *Rhetoric of Fiction* (1961) with four general rules, and 'Types of Narration', 'Dramatised Narrators', 'Reliable Narrators', and of course also 'Unreliable Narrators' – not to speak of his 'Control of Distance', and his 'Control of Side Views' etc.? Many more could be quoted.[16]

Plato finally in the *Phaedrus* gives the right principles for argumentative writing, and his remarks cease to apply to poetics. Plato maintains that a good argument should be based on truth, not on illusion or trickery, and truth is reached through the rational processes of analysis and synthesis, or, as Plato calls them here, *diœresis* and *synagogé* (265 D-E). Synthesis provides the general definition of the subject to be discussed; analysis draws the divisions of the argument from the definition so provided. The two processes make up what Plato here calls dialectic, and he adds that the whole of rhetoric is contained in them (269 B). A similar argument could be made for poetics: the divisions of the poem are determined by its subject, or by the configuration of the poetic image; but Plato does not carry the method into poetics. Later Longinus will do it in his *Sublime* (ch. 10, on Sappho's Ode).

Plato refers again in another dialogue to the principle of organic unity, but avoiding the organic metaphor (which incidentally confirms that the metaphor is not indispensable to the principle) and he resorts to a different simile. In the *Gorgias* (503 E-504 A) he says:

> The orator, like other craftsmen (*demiourgoí*), has his own particular work in view and thus selects the things he requires for that work, not at random, but with the purpose of giving a certain form (*eídos*) to whatever he is working upon. You have only to look, for example, at the painters (*zógraphous*), the builders, the shipwrights, or any other craftsmen, to see how each of them arranges everything according to a certain order (*táxis*) and forces one part to fit with another, until he combines the whole into a regular and well-ordered production.

In these discussions an idea has been frequently appearing, that of the *whole*, which will play an increasing role in later speculation. In Plato's *Charmides* (156 C-E) it is already applied to medicine. A doctor cannot cure, for example, the head of the patient alone, he must also cure the rest of the body; he shoud treat 'the whole and part together' for 'the part can never be well unless the whole is well'. Still another metaphor is used in the *Republic*, that of the work of art, such as a statue. The artist should 'consider whether, by giving this or that feature its due proportion, we make the whole beautiful' (420 D). That principle is then applied to the plan of the ideal State. In the *Theaetetus* (204 A) the principle is advanced that 'the whole made up of the parts is a single notion (*eídos*) different from all the parts'. This passage seems to be the origin of the often repeated maxim that the whole is more than the sum of the parts.[17]

To conclude, Plato made an important contribution to aesthetics in the *Phaedrus* when he enunciated the principle of the organic unity of a composition, which was to become the keystone of later systems of criticism. Plato definitely affirmed its value for the judgment of poetry, and not only for oratory, as has been thought. Furthermore, he also considered, as was his wont, the philosophical principles involved in it. I have attempted elsewhere to show succinctly the role of the principle in later thought up to the present day. As I said, using mathematical language, organic unity, if a *necessary* condition for the composition of a work of art, is not *sufficient* for it. Other factors are called for, such as imagination, feeling and taste. Without them, we may obtain a well-constructed work of the intellect or of the practical reason, but not a work of art.

Notes

1 H. De Witt Parker, *The Analysis of Art*, New Haven, 1926, p. 37.
2 J. Hospers, *Meaning and Truth in the Arts*, Chapel Hill, 1946, p. 10.
3 J. Stolnitz, *Aesthetics and Philosophy of Art Criticism*, Boston, 1960, pp. 232–3.
4 E. Panofsky, *The Life and Art of Albrecht Dürer*, 4th edition, Princeton, 1955, pp. 261 and 276.
5 S. H. Butcher, *Harvard Lectures on Greek Subjects*, London, 1904, p. 192.
6 A. Trendelenburg, *Logische Untersuchungen*, 3 Aufl., Leipzig, 1870, pp. 143–7.
7 R. Eucken, *Main Currents of Modern Thought*, English translation, London, 1913, pp. 165 ff.
8 Daniel Stempel, in 'Coleridge and organic form: the English tradition', *Studies in Literature*, VI, 1962, p. 93, traces Coleridge's definition of the organism back to Hume, who gave a definition of it before Kant, in his *Treatise* of 1739. Hume says that the parts of an organism exhibit 'the reciprocal relation of cause and effect in all their actions and operations' (Part IV, Section vi, ed. Selby-Bigge, p. 257). Now the significance of the terms 'cause and effect', instead of Kant's later 'means and end', lies in the fact that the concept of cause is not incompatible with a purely mechanistic explanation of life, while 'means and end' is, being teleological. Even more explicitly mechanistic is Kames's description of the organism in 1762, which Stempel also quotes in support of the preferability of 'English sources' for Coleridge's formula (p. 90). Kames repeatedly describes the living organism as a 'mechanism'; of 'a wonderful subtility', but still a mechanism.

In any case, Coleridge's formula repeats Kant's, and not Hume's. If Coleridge was aware of Hume's formulation, he rejected it and preferred Kant's. As an English scholar said on a similar occasion, 'there is no need to apologise for introducing a German humanist into the consideration of English literature. In studying the 19th century we cannot become insular. During that period culture gradually became so cosmopolitan (either by instinct or imitation) that any author may be regarded as English who illustrates or influences English literature' (H. V. Routh, *Towards the 20th Century: Essays in the spiritual history of the 19th century*, Cambridge, 1937, p. 231).

9 See 'The organic concepts in aesthetics', *Comparative Literature*, XXI (1969), pp. 1–30; also, an article on 'Organic unity' in the *Dictionary of the History of Ideas*, New York, Scribners, 1972. A briefer account, in Italian, with some additional references, appears in *Critica e storia letteraria*, Studi offerti a Mario Tubini, Padua, Liviana, 1970, pp. 30–6.

10 R. Hackforth's comment in his *Plato's Phaedrus*, Cambridge, 1952, p. 130, shows the significance of the aesthetic reference gradually growing.

11 See A. Boeck, *Enzyklopädie und Methodologie der philologischen Wissenschaften*, Leipzig, 1877, p. 133; Butcher's 1904 remarks, quoted in n. 5 above; J. B. H. Atkins, *Literary Criticism in Antiquity, A Sketch of its Development*, London, 1952 (1934), I, pp. 54–5; W. Jaeger, *Paideia*, III (1934), ch. 8 (p. 329 of Italian translation, Florence, 1959); H. Cairns, Introduction to *Lectures in Criticism*, Johns Hopkins University, New York, 1949, p. 5; P. Vicaire, in the fullest account of Plato's ideas on literature (*Platon, critique, littéraire*, Paris, 1960), signals the appearance of the organic concept and its aesthetic significance, pp. 287 and 360; see also his subsidiary work, *Recherches sur les mots designant la poésie et le poète dans l'oeuvre de Platon*, ibid., 1964; while H. Osborne (*Aesthetics and Art Theory, An Historical Introduction*, New York, 1970, p. 285) considers the *Phaedrus* doctrine only as a contribution to rhetoric, Vicaire points out that there the term 'rhetoric' acquires a negative denotation, 'ne designant que la vaine virtuosité' (p. 287), as common in modern times; but Osborne makes use of the organic concept in his own system of aesthetics: see my paper of 1969, quoted in n. 9, p. 22; also, G. M. A. Grube, *The Greek and Roman Critics*, London, 1965, pp. 46 and 58–60; F. S. Dorsch, *Classical Literary Criticism*, Baltimore, 1965 (Pelican Series), p. 13.

12 Grube, op. cit., p. 57 n.

13 266 D – 267 D, adapted from Jowett's and Hackforth's translations.

14 See my paper 'Coleridge and Schlegel reconsidered', *Comparative Literature*, XVI (1964), pp. 97–118.

15 N. Lemercier, *Cours analytique de littérature générale*, Paris, 1817, 4 vols.

16 Cf. Henri Peyre: 'The ridiculous and scholastic vogue for rhetoric and for figures of speech and for categories or "modes," which seems to revive the worst features of Greek and Latin grammarians, keeps many sophisticated students in America from feeling freshly and thinking independently.' *The Failures of Criticism*, Cornell University Press, 1967, p. 309 n. 1.

17 Plato dealt with the general concept of unity, particularly in the

Parmenides, 137 C. So did Aristotle in *Metaphysics*, Book Delta, chapters 6 and 16; his concept of unity, in general, and connection with art, was recently discussed by H. Osborne, op. cit. in n. 11 above, pp. 285–6. For Clement of Alexandria see E. F. Osborn, *The Philosophy of C. of Alexandria*, Cambridge, 1957, p. 17. The Neoplatonists carried on the speculation on unity in general, e.g. Plotinus, *Enn.*, VI, vi, 2, and Proclus, *Elements of Theology*, LXIX.

Philip C. Ritterbush

Organic form: aesthetics and objectivity in the study of form in the life sciences

The degree of objectivity attained in the depiction of plants and animals in the European visual arts by the end of the thirteenth century permits confident identification of the species portrayed in the decorative arts of carved ornament and the illumination of manuscripts.[1] Two centuries later, works by Pisanello, Mantegna, Botticini, and Dürer exemplified the establishment of standards of realism in the artistic treatment of living nature at a time when the descriptions and illustrations in herbals and zoological treatises were still very imperfect.

Leonardo da Vinci (1452–1519) considered naturalistic depiction to evince a superior mode of knowing. Verbal statements could convey 'but little perception of the true shapes of things'.[2] He maintained that an artist could represent essential aspects of organisms by showing the changes through which they develop. By knowing how to perceive nature (*saper vedere*), the artist might attain the highest powers of knowledge. By the middle of the sixteenth century accomplished and accurate drawings became regular features in treatises on anatomy, as in *De Humani Corporis Fabrica* by Vesalius (1543), zoology, as in Conrad Gesner's *Historia Animalium* (1551), and botany, exemplified by Leonhart Fuchs's *De Historia Stirpium* (1542). Leonardo's studies of growing plants, storms, and waterfalls are especially clear examples of 'the dynamic tendencies of the Italian Renaissance where', in Erwin Panofsky's words, 'all things, whether alive or inanimate, were interpreted as organic entities molded and stirred by inherent forces'.[3]

The standards of pictorial realism became institutionalized in the natural history collections and descriptive monographs of the seventeenth and eighteenth centuries.[4] By the middle of the eighteenth century directories of plants and animals by Carl Linnaeus (1707–78) applied powers of description and capacity for drawing fine distinctions to the extensive enterprise of preparing scientific inventories of the natural world. Linnaeus claimed to have examined directly all known species of plants. He was a poetically sensitive observer who enthusiastically led caravans of students into the Swedish countryside and responded ecstatically to discoveries of the sexuality and reaction movements of plants.[5] The dynamic aspect of plant life interested him greatly. He thought that each region of the cross-section of the stem developed progressively into a different

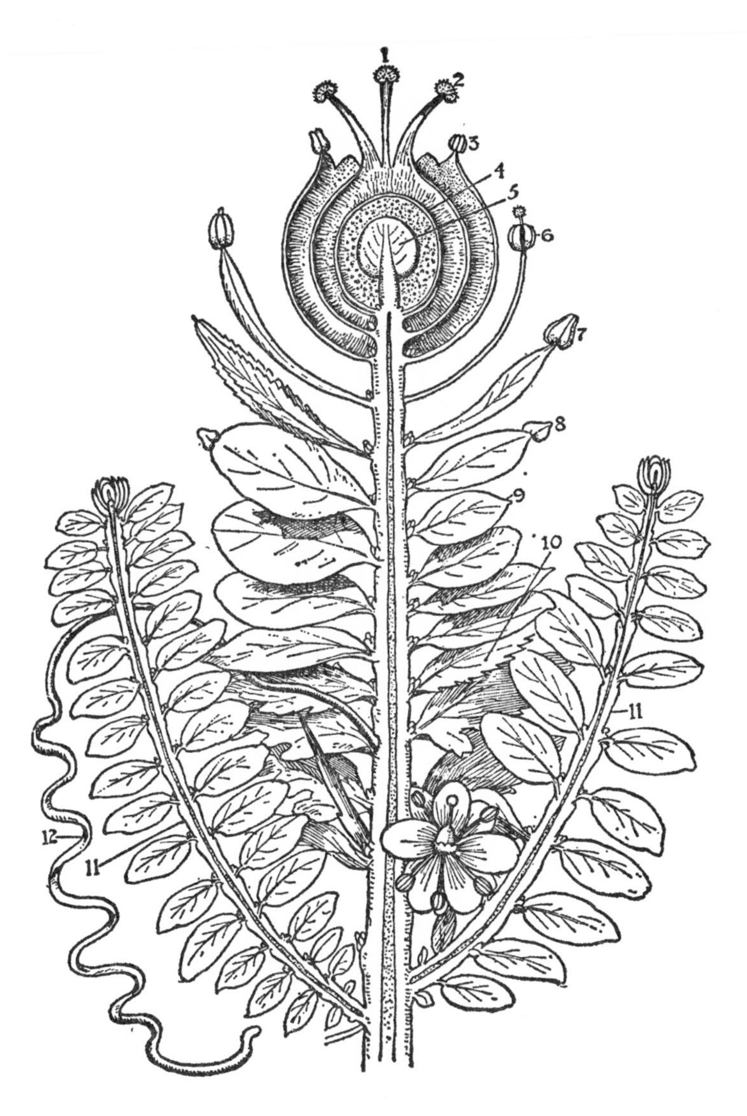

Figure 1. Part of Goethe's figure of the variations of leaf form according to which he interpreted the parts of the flower: 1 and 2 the stigmas of carpels; 3 and 4 walls of the seed-box; 5 the seed leaves or cotyledons; 6 typical stamen; 7 slightly petaloid stamen; 8 intermediate between stamen and petal; 9 petal; 10 sepal; 11 compound leaf with pinnules; 12 a leaf transformed into a tendril.

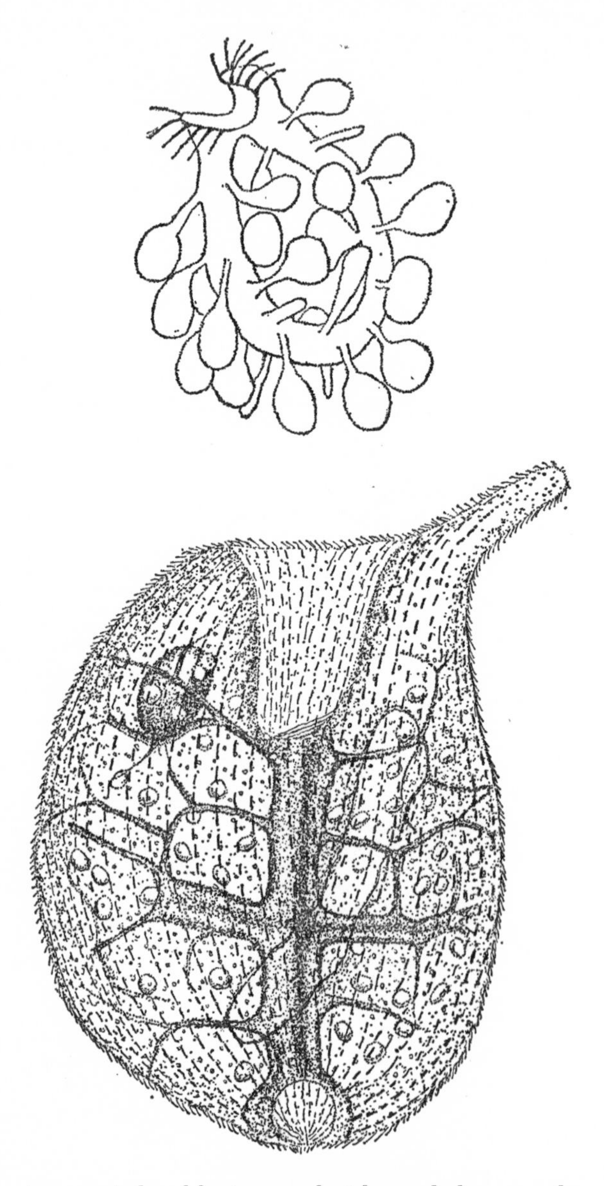

Figure 2. An error induced by transcendental morphology was the portrayal of stomachs in single-celled organisms which do not possess such structures.

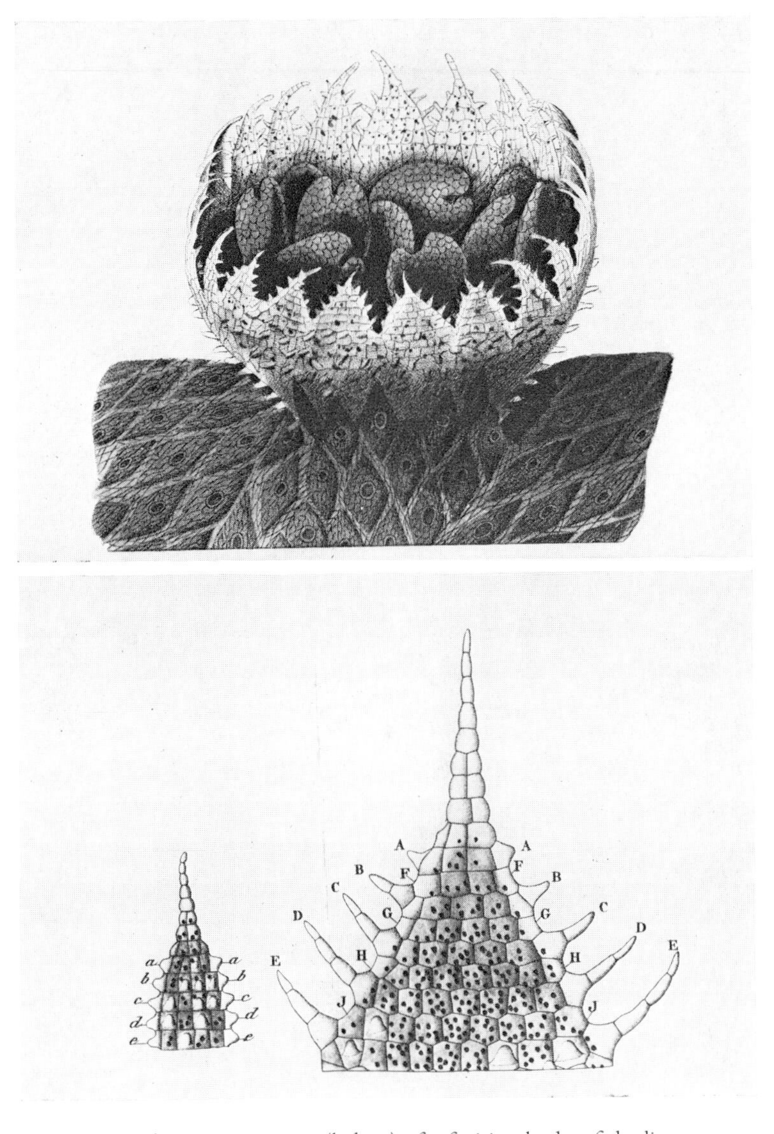

Figure 3 In the cusp structures (below) of a fruiting body of the liverwort *Marchantia polymorpha* (above), C. F. Brisseau-Mirbel observed that growth occurred by the division of cells, so that as the lines of cells marked a, b, c, etc. in the left-hand figure lengthened and grew into lines marked A, B, C, etc. in the right-hand figure, there appeared between them new lines of cells F, G, H, and J. I believe this constituted the discovery that the growth of organisms takes place through cell division.

Figure 4 The seed bud of *Passiflora*, cross-sections in which arrays of cells arranged around the fertilised ovum reveal successive stages of growth.

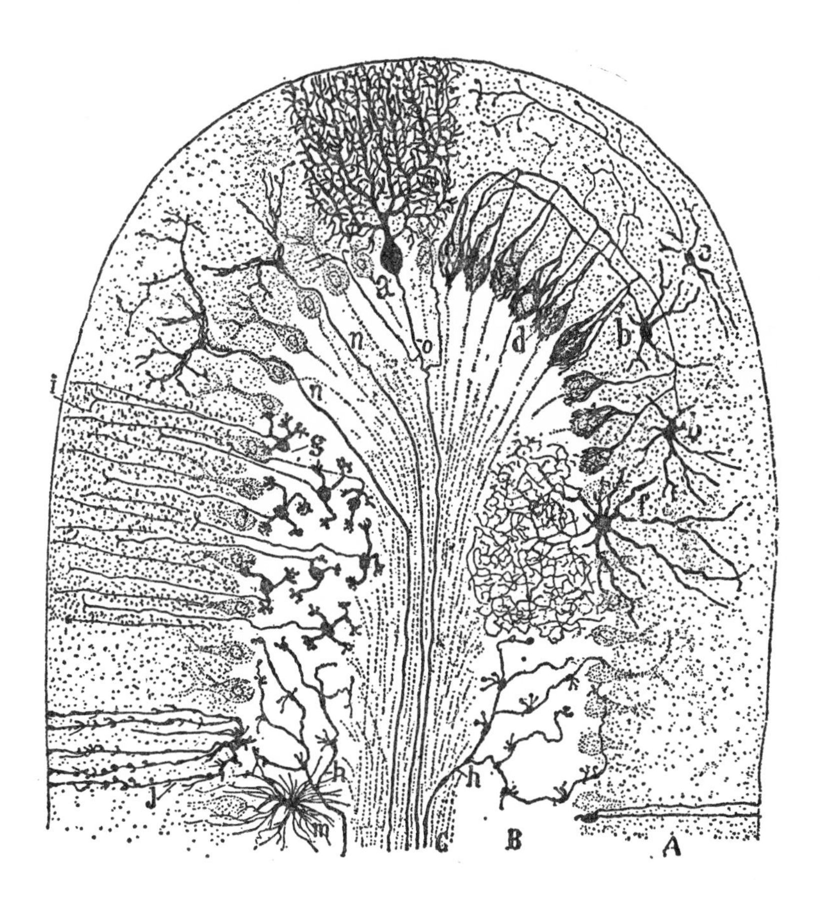

Figure 5. A schematized cross-section of a convolution in the mammalian cerebellum showing different cell types, drawn by Ramón y Cajal in 1894. 'The garden of neurology holds out to the investigator captivating spectacles and incomparable artistic emotions. In it, my aesthetic instincts found full satisfaction at last.'

part of the flower by a process he styled the metamorphosis of plants.[6] Linnaeus praised the beauty of plants with an exuberance that repeatedly breaks forth everywhere in his voluminous writings.

Rousseau became an ardent disciple of Linnaeus and a devoted collector of plants. Like him he experienced feelings of 'ecstasy' and 'rapture' at the 'thousand little acts of fructification I observed'.[7] Everywhere in Europe fashionable amateurs reacted effusively. Writing about observations that plants responded sensitively to external stimuli, Hannah More found this quality 'so pretty and poetical' that she proposed to write 'the tragedy of Flora, with the dramatis personae from the parterre and the chorus from the shrubbery'.[8]

Erasmus Darwin (1731–1802), grandfather of Charles Darwin, composed a versified scientific epic, *The Loves of the Plants* (1789), describing reproduction in some eighty different kinds of plants, each assigned to its proper class in the Linnaean system according to the number of stamens. It was a mild form of pornography employed in the cause of popularizing scientific knowledge. Typical of the personified descriptions is this account of the honeysuckle:

> Fair LONICERA prints the dewy lawn,
> And decks with brighter blush the vermil dawn;
> Winds round the shadowy rocks, and pancied vales,
> And scents with sweeter breath the summer-gales.
> With artless grace and native ease she charms,
> And bears the horns of plenty in her arms.
> *Five* rival swains their tender cares unfold,
> And watch with eye askance the treasured gold.

A note explains that five males, the stamens, encircle the female organ of the flower, at the center of a tubular extension of the petals, shaped like a horn of plenty, requiring any insect that visits the flowers to have a 'long and pliant proboscis'.[9]

Darwin hoped that the visual character of his descriptions, accented vividly with engravings, some by William Blake, would enable the poem to reach a wide audience, who would be led through the notes to understand a wide range of topics in biology, anthropology, geology, chemistry, and other fields. A very generous appreciation of the significance of his writings and his influence upon the

English Romantic poets has recently been offered by Desmond King-Hele.[10] Darwin's dedication to wider understanding of science shows up in his belief that priestly élites had sought to monopolize scientific knowledge by coding important insights as myths, that, for example, the fable of Arachne was a code for the arts of weaving and that the marriage of Pluto and Persephone symbolized the interrelations of matter and energy.[11] The response to Linnaeus seemed to confirm Darwin's thought that an appreciation of natural beauty might be the means by which scientific concepts might be communicated to a large audience.[12]

In a summary treatise about the principles of botany Linnaeus took account of the various ways that leaves are arranged around the plant stem: whorled; arranged as clusters of three, four, five, etc.; opposed; alternate; dispersed; connected; overlapping; in bundles; or in opposed rows.[13] Any of these arrangements might be cited as a character by a botanist describing a species of plant. They are important attributes of plants, presumably constant with each species, yet represent only one among many descriptive criteria and so received little attention from systematic botanists. But in 1754 the Swiss naturalist Charles Bonnet (1720–93) and the mathematician Calandrini of Geneva reached a radically advanced understanding of these different arrangements by perceiving that the patterns of leaf arrangement were generally quite limited and that in 125 species only five basic orders of arrangement were found: alternate, opposed, in clusters, quincuncial (each leaf in spirally arranged groups of five occupies the same position on the stem as the corresponding leaf in another group of five), and spiral.[14] Bonnet noted the advantage to plants in exposing successive leaves to light and air, thus relating arrangement to the functions of transpiration of water and exchange of gases. He also understood the categories of arrangement to reflect different orders of symmetry around an axis. With this insight the study of organic form may be said to have begun, going beyond the descriptive statement of a character or the direct illustration of a specimen to analyse the growth of organisms in terms of governing principles.

Bonnet's conception went well beyond the discrimination of observable differences. He analysed the patterns of leaf arrangement in terms of a process so basic that only a limited number of variations were possible – the extension of growth along an axis with leaves appearing either singly in series or on two or more sides of the stem at once; either spiral, reflective, or radial symmetry. This constituted a discovery that principles of spatial arrangement might be abstracted from organisms as aids to understanding vital process. As such it marked the beginning of the scientific study of form.

As used in biology, form is a quality derived from the analysis of an object whereby its underlying structure is represented. It is not a chance projection like a shadow but is congruent with its object in constant and lawful ways. It is subject to empirical verification by measuring the object. Most importantly, it represents the processes governing the achieved structure of the plant or animal. Bonnet's contribution was to analyse the leaves and stems of plants so as to yield an important principle of form. Although he did not develop his insight into a general concept of form or even use the term, his treatment represents an important landmark in the development of biological method.

Nothing in Bonnet's discussion indicates that the manner of growth of plants sets them off from non-living nature. It was quite usual for writers to compare the growth of plants to the growth of crystals or mineral substances. The *arbor Dianae* – dendrites of silver – was often cited as a model for plant growth. Fossil plants and branching corals were commonly conceived to have developed through processes identical to those governing the growth of crystals. Snowflakes and moss agates indicated that 'nature can and does work the shapes of plants and animals without the help of a vegetative soul'.[15] The pioneer plant anatomist Nehemiah Grew (1628–1712) supposed that the units of matter existed in the form of four basic crystal shapes which combined in various ways to form branching structures, curved vessels, or fibres.[16] According to the assertive mechanical philosophy of the seventeenth century the growth of living things took place through simple aggregation – like building a wall of bricks. Lacking means of direct observation of processes of nutrition and growth, writers of the time supposed they took place in the simplest imaginable way. The Cartesian concept of particulate matter ruled out any essential difference between the form of organisms and non-living entities. By 1801 it was understood that crystals were mathematically more regular than organisms, after Haüy demonstrated that crystal form was governed by strict geometrical principles.[17] But it was quite usual for eighteenth-century naturalists to refer to plant and mineral growth interchangeably.

It was among poets and critics, well before analytic scientific standards developed, that the special character of organic form was first appreciated. Edward Young's *Conjectures on Original Composition* (1759) pointed to a key attribute of the form of living things in praising the originality of great compositions. 'An *original* may be said to be of a *vegetable* nature; it rises spontaneously from the vital root of genius; it grows, it is not made.'[18] Young was contrasting the outward aspect of vegetation – tendrils, sprouting seeds, and curving surfaces – to more mechanical processes of aggregation as in crystals or machines, suggesting that creative thought might be set above routine association of ideas. The image of responsiveness and complexity offered by plant growth was an appealing contrast to the simplistic portrayal of mental activity offered by materialistic psychologists, who represented thought as the addition of one element to another.

In an historical study of literary concepts of creativity Professor Meyer Abrams points to this image as the signal of a revised appreciation of the intellectual processes which flower in a work of art.[19] If one sees the association of ideas as a repetitive, mechanical process and looks toward the redefinition of imagination as an active force shaping the raw material of thought, then the introduction of the image of the growing plant marks the transition from one view to the next. Within the context of the history of science the transition is less clearly marked because the extent of the difference in form between the living and non-living had not yet been recognized. As life was understood in the middle of the eighteenth century plants might seem to consume more energy (in ways as yet poorly defined) than minerals, to require more nutrients for their growth, and to stand higher in the scale of life, but the properties of form distinguishing them from the non-living had not yet been defined. A plant might serve, as it had for Leonardo, as an emblem of the dynamic processes shaping the world, including movements of the land, winds, and flowing water, as well as the restless urges of living things. So Abrams puts it, that 'The imagination, in creating poetry, therefore echoes the creative principle underlying the universe.'[20]

I do not wish to appear to argue with his interpretation that a more generous estimate of the power of the imagination is reflected in the use of the plant form to signify creativity, for I believe this was indeed so, as inadequate mechanistic conceptions of mental activity came to be succeeded by recognitions of the dynamic qualities of creative thought.[21]

What Abrams seems not to have considered is the possibility that associating plant forms with creative mental activity may have played a decisive part in establishing not only a new view of artistic creativity but also a new conception of plant life and form. Artistic achievements paved the way toward naturalism in the description of plants and animals by naturalists. In a somewhat similar way literary concepts helped to win recognition for the properties that distinguish the form of organisms from the non-living. J. W. von Goethe (1749–1832) had frequently noticed that individuals of a given species of plant were to be found growing to different sizes or in varying forms, depending on whether they occurred in mountains or lowlands, in light or in shade. He was one of the first students of environmental influences on plants, an aspect of nature that occupies a central position among topics for modern biological research. Yet in his time there was barely any competent understanding of the influence of such factors as light, heat, day length, or gravity upon plant growth, and too little knowledge about the material processes of growth to serve as a basis for sound conjecture about the meaning of his observations. This did not prove to be a serious obstacle to his genius and he set out to formulate a theory of plant life that would do justice to his powers of intuition and express his apprehensions about the central phenomena of growth and form.

Acutely aware of the diversity of shape and structure in plants and longing for some principle by which his observations of the previous ten years could be rationalized, Goethe left Weimar in 1786 for a two-year tour of Italy. As he wrote to Frau von Stein before leaving, the plant kingdom was raging in his mind. He felt that he stood on the brink of a great discovery. 'And it is no dream, no fantasy, it is the awareness of the essential form with which Nature is, as it were, always toying and in the course of play brings forth the infinite variety of life.'[22] At the botanical garden in Padua he was overwhelmed by the beauty and variety of the plants he beheld. If only he could discover for all plants a single source to which any form might be traced. He did not mean an ultimate ancestor of plants in an evolutionary sense, but rather gave it as his belief that there was an ideal type from which all other plants could be derived intellectually. The *Urpflanze*, as he styled it, would express the essence of vegetative life directly to its beholder. It would typify all of the different kinds of plant life. He expected to know it at once if he came upon it. In April of 1787 in the Public Gardens at Palermo, while musing upon the plot of a play to be based on the Odyssey, he had these thoughts.[23]

> Seeing so much new and burgeoning growth, I came back to my old notion and wondered whether I might not chance upon my archetypal plant. There must be such a plant, after all. If all plants were not molded on one pattern, how could I recognize that they *are* plants? . . . My fine poetic resolutions were frustrated. The garden of Alcinous had vanished. In its place the garden of the world opened up. Why are we moderns so distraught? Why are we challenged to demands we can neither attain nor fulfill?

A month later he wrote from Rome to his long-standing friend, the poet and writer on human nature, J. G. Herder, that the archetypal plant 'will be the strangest growth the world has ever seen'. 'Nature herself will envy me for it.' From such an image an infinite range of plant types could be projected. 'They will be imbued with inner truth and necessity. And the same law will be applicable to all that lives.'

Goethe did not succeed in finding the primal plant. Rather he altered his idea of it, coming to believe that all plants were composed of modified leaves, except for the stem, considered as the geometric axis upon which the variations of growth were exhibited. Thus he abstracted from all plants a basic element, the primordial leaf, which might be expressed as a leaf, a flower part, or a seed capsule. This process of growth through the repetitive elaboration of a basic plan he conceived to depend upon cycles of expansion and contraction whereby organs were formed according to the characteristic rhythm of each plant. After two years' further development of these ideas Goethe incorporated them into his treatise of 1790, 'An Attempt by

J. W. von Goethe, Privy Councilor of the Duchy of Saxe-Weimar, to Explain the Metamorphosis of Plants'. In 1799 he expressed his ideas in the form of a poem which included these lines:

> Oft the beholder marvels at the wealth
> Of shape and structure shown in succulent surface –
> The infinite freedom of the growing leaf.
> Yet nature bids a halt; her mighty hands,
> Gently directing even higher perfection,
> Narrow the vessels, moderate the sap;
> And soon the form exhibits subtle change.

In 1795 Goethe published a paper summarizing rather similar views about the comparative anatomy of vertebrates. There also he sought an ideal type from which all skeletal forms might be derived. To him the animal skeleton was a series of vertebrae, no one of which could attain maximum development without proportional reduction in the others. He returned again and again to this interest in comparative anatomy, always in terms of units whose combinations and modifications revealed an ideal plan for organic life.

We must take particular note of the visual character of the idealized plant form. It exists not in nature but in the mind, which it delights with all the vividness of a dream. Goethe sought to understand natural processes by visualizing them. Throughout his works we find a preoccupation with visual experience which was undoubtedly influenced by his youthful ambition to become an artist and his early efforts at painting and drawing. One of his principal recreations was to view aquatic life through the microscope. In 1786 he made detailed drawings of protozoa, including paramecium and vorticella. He eagerly corresponded with Eduard Joseph d'Alton (1772–1840) about scientific illustration. Goethe supposed that he had fathomed the essence of phenomena when he had worked out a means of representing them visually as a chart or diagram, a portrait of organic nature. Thus he had the warmest praise for a chart published in 1821 showing the distribution of life in the world's oceans, correlated with climatic factors. He regarded it as a highly successful attempt 'to present to the eye, by symbolic means, facts which have a sensuous basis yet are not visibly perceptible, so that imagination, memory, and understanding may be stimulated to fill in what is

missing.'[24] During his Italian journey, for example, he investigated technical aspects of painting, carefully studying the effects produced by the use of different colors and questioning painters about how they worked. He had Angelica Kauffmann paint a variety of landscapes from which the usual colors were omitted so as to compare them to the atmospheric effects he observed in nature. A flower-painter might attain accuracy in depicting nature through mere imitation, Goethe wrote, but the highest artistic expression would require a profound understanding of the flower, its mode of growth, and the effects of the environment, so that its inmost essence could be expressed visually.[25]

Biologists refer to the study of form as 'morphology', which differs from anatomy in seeking to elucidate the processes governing achieved structure rather than describe the structure itself. Goethe was among the first to use the term. His *Metamorphosis of Plants* was an influential early publication in the field. What prevented that work from contributing to morphological knowledge was its lack of concern for the material processes by which growth took place, such as the different structures of the various tissues as they might reveal direction or rate of growth. He totally ignored the properties of the substances composing the organism and the processes by which these contributed to its growth. That leaves, petals, and sepals were variations upon a basic element in development had already been proposed by Linnaeus. Goethe gives an account of form only in its ideal aspects. The abstraction of the primordial leaf from the varied aspects of the plant explains nothing about the manner in which growth actually occurs, but it is the supreme example of Goethe's desire to project within his own mind a visual impression of the workings of nature. He found it thrilling to apply his abstraction to plants and it satisfied him more than any scientific explanation that his contemporaries had provided.

The aspects of form that so excited Goethe were divorced from the reality they were intended to represent. Pursuing his quest for visual treasure he was led further and further into realms of illusion, as though despairing of any possibility that the properties he sought might exist objectively among living things. In spite of his professions of love for the appearances of nature, beauty remained for Goethe a neoclassic ideal property. His visual image of the ideal plant lacked any scientifically meaningful relation to the processes of growth it was intended to represent. The elements of the image could not be matched with the parts of plants as they might have been observed or the processes by which they grew. His notion that the skull was composed of degenerate or fused vertebrae could not have withstood direct examination of a series of animal skeletons. His schematizations were profoundly satisfying conceptually, as his enthusiasm for their translation into poetry showed. His attempts to portray nature were not mistaken simply because they contained errors of fact but because they were not adequately grounded in observation. The great neurophysiologist Sir Charles Scott Sherrington, who himself wrote poetry and devoted great care to working out his general ideas of nature, delivered in 1942 a Deneke lecture at Oxford on Goethe's view of nature. It is to this idea of nature rather than to organisms themselves that Goethe's images refer. Nature is fecund and powerful, constantly changing, symbolized by a pantheon of deities ceaselessly reshaping the world. 'Nature resembled not too distantly a vast Brocken-scene in which the supernatural worked the natural, with Faust as spectator.'[26] Goethe's nature is orphic after the fashion of the Greek mysteries. Man's greatest privilege is to be initiated into a sense of her awesome beauty. Through her votary the poet, Nature becomes known, as in the *Urworte* poems of 1817, which appeared in the second volume of his writings on morphology.[27]

The concept of form employed by Goethe had a number of serious weaknesses which may account for the illusory character of the primal phenomena he sought. He did not adequately conceive the relation of part to whole in organic form. He erred in supposing that if the form of an organism does not arise from the mechanical aggregation of parts it must result from the operation of a transcendent principle. The truth lay between these poles, as was convincingly argued by Denis Diderot (1713–84) in *D'Alembert's Dream*, dialogues written in 1769 but not published until 1830. He maintained that properties of organisms result from the manner of arrangement of their constituent parts. He compared a single organism to a swarm of bees whose form changed as its units moved about. He strengthened the analogy to the organism by pointing out that such a colony might be transformed into a single animal if the units were somehow fused together.[28] The form of any organism is an emergent property of its structural elements. Sensitivity and other faculties of the organism as a whole are destroyed if it is broken into its parts. 'Everything conspires to produce the kind of unity that exists only in a living animal.'[29] Diderot recognized that order emerges as elements are assembled into a living whole, as a derivative of the collective character of vital processes, so that life consists of an interplay between the properties of simple substances and the complex organizations according to which they are arranged. This solved a key problem confronting the critics of simplistic Cartesianism. The form of the organism did not result from the arrangement of simple building blocks but was an expression of tensions or complexities inherent in the very number of building blocks required. The whole is more than the sum of its parts.

Goethe also insisted on interpreting the development of organisms as strivings to fulfil an ideal design. He failed to perceive the advantages of the conception offered by Immanuel Kant (1724–1804), who maintained in the *Critique of Judgement* (1790) that the living organism emerged as a result of the shared capacity of its parts, not as the fulfilment of an end in view guiding a process of assembly. It grows. It is not planned. It exhibits purposiveness without a governing end: *Zweckmässigkeit ohne Zweck*. Goethe's appetite for beauty led him astray on this point because he presupposed a design. Kant held instead that our sense of the beauty of nature is grounded in our aesthetic faculties and need not imply an ideal or pre-existing design. If a natural form exhibits beauty it is because of its harmony with the patterning faculties of human consciousness. The experience of beauty derives from a resonance with nature, man's response to the world which shaped him and conditioned his perception. Nature offers, in Professor Wimsatt's words, 'a highly satisfying fitting of experience precisely to our faculty of experiencing, to the progress of our knowledge'.[30] The regularities perceived in nature give rise to aesthetic feelings of pleasure which are auspicious for knowledge. The perception of beauty gives a foretaste of knowledge because the faculties by which we understand the world of nature derive from our responsiveness to its regularities and rhythms. We are most likely to be satisfied by that which we are constituted to perceive. Thus all accounts of natural form would necessarily include a subjective factor upon which depends our ability to recognize pattern and coherence in living systems, without having to postulate a designer. This insight nicely complemented Diderot's recognition that order might be an emergent property of complex living systems.

I do not believe that Goethe properly appreciated the relation of form to the organism itself, a relation that Charles Bonnet had properly understood, or the way that organic form was defined by the interactions of systems of component parts without any necessity for an archetype or ideal. Ernst Cassirer is undoubtedly right to recognize a bond of sympathy between Kant and Goethe, deriving from their common interest in questions of form, but I do not believe that Goethe's idealized conception of form responded to Kant's understanding of organic development.[31] Goethe fully appreciated the exalted role of the sense of sight in understanding life processes and clearly appreciated the indebtedness of scientific knowledge to aesthetic aspiration. In his rhapsodic stance of admiration before nature he was frozen in the posture of Linnaeus of two generations before. Despite the vastness of his creative accomplishment the elements that would most decisively influence the future of the understanding of organisms were those he had passed over. For their integration into a synthesis opening a new chapter in the history of biology we must look to Coleridge.

The Ancient Mariner (1798) reaches its climax at the moment of the mariner's vivid awareness of beauty in the 'rich attire' of the water-snakes and elegance in their tracks of 'golden fire', where before he had been repelled by the appearance of the 'million million slimy things'. The recognition of beauty is perhaps the central symbolic occurrence in Romantic poetry, as the greatest achievement of that literary generation was its analysis of the experience of beauty. In this endeavor Coleridge played a leading role and organic form figures as largely in his conception of beauty in the arts as it had in the central event of his poem.

In a number of lectures and essays, of which the most important was 'On the Definitions of Life' (1830), Coleridge criticized empirical physiologists as pursuing too narrow an approach to the problems of life. They studied organisms as though they were simple physico-mechanical assemblages, a piece at a time, while the most important phenomena of life resided at the level of the organism as a whole and could be understood only through the exercise of comprehensive powers of insight. While Coleridge did not propose an extra-scientific speculative approach, he readily saw the inadequacies of a stepwise analytic method. A machine could best be understood by taking it apart, but the dissection of an organism yielded members whose meaning was lost when they were removed from the intact body. Only the elementary powers of the mind were required to design a machine or to analyse the structure of a crystal. But to understand the organism the mind must project something of itself into nature, perceive with its highest integrative powers, and reshape the raw sensations of experience. This, in Coleridge's celebrated if somewhat obscure conception, was the faculty of imagination.[32]

Coleridge maintained that the imagination was an active shaping faculty. He exalted its creativity far above the routine processing of elements of information. The powers of imagination transcended those of the mere association of ideas just as the living organism showed higher powers than those required for the aggregation of molecular units into crystals. The creative imagination finds, in Coleridge's words, 'correspondences and symbols' in the growth of a plant, as it changes nutritive substances into its own substance during growth. Before a living plant, he wrote, 'I feel an awe, as if there were before my eyes the same power as that of the reason.'

Coleridge held that the true work of art was that which displayed the characteristics of an organism rather than a mechanical device. He restated Schlegel's distinction between mechanical and organic form:[33]

> The form is mechanic when on any given material we impress a pre-determined form, not necessarily arising out of the properties of the material, as when to a mass of wet clay we give whatever shape we wish it to retain when hardened. The organic form, on the other hand, is innate; it shapes as it develops itself from within, and the fullness of its development is one and the same with the perfection of its outward form. Such is the life, such the form. Nature, the prime genial artist, inexhaustible in diverse powers, is equally inexhaustible in

> forms. Each exterior is the physiognomy of the being within, its true image reflected and thrown out from the concave mirror.

Organic forms were defined by five attributes repeatedly stated by Coleridge. Of great consequence for the future, they are well worth noting, for their influence extended far beyond the realm of aesthetics as traditionally received. First, the origin of the whole precedes the differentiation of the parts. The whole is primary. The parts are derived. In the organic form 'the whole is everything and the parts are nothing.'[34] 'Whatever is truly organic and living, the whole is prior to the parts.'[35] Second, an organic form conveys the process of its own development to the observer, manifesting itself as the end product of a sequence of development whose course may be inferred from rings of growth or layers of deposited shell. 'Productivity' or growth is the first power of living things, and it exhibits itself as 'evolution and extension in the Plant'.[36] Third, as it grows the organism assimilates diverse elements into its own substance:[37]

> Events and images, the lively and spirit-stirring machinery of the external world, are like light, and air, and moisture, to the seed of the Mind, which would else rot and perish. In all processes of mental evolution the objects of the senses must stimulate the Mind; and the Mind must in turn assimilate and digest the food which it thus receives from without.

Fourth, the achieved form of the plant is directed from within, as Coleridge observed in the quotation above. The external aspect of living things is determined by internal processes, not, as in a human artefact, from without. Fifth, the parts of the living whole are interdependent. As Professor Meyer Abrams puts it, 'Imaginative unity is an *organic* unity: a self-evolved system, constituted by a living interdependence of parts, whose identity cannot survive their removal from the whole.'

Coleridge developed this concept of organic form out of an analysis of Shakespeare's plays, which he held to have been designed from a unified view of human nature. Each play was deemed organic because part and whole were interdependent and informed by a single conception of man's place in nature, and also because the subject matter was well adapted to the dramatic purpose. Having found these formal qualities, which satisfied the criteria for organic form, Coleridge praised the work of art manifesting them as a genuine product of the imagination. On just the same grounds Friedrich Schlegel had praised Goethe's novel *Wilhelm Meister* as a perfectly organic work of art, wherein the parts repeated the whole and all elements were interdependent.[38]

Goethe's striving after universal forms was not without influence on biology, although it failed to promote meaningful research. His devotion to ideal forms influenced a whole school of transcendental morphology. Lorenz Oken (1779–1851) considered every organism to be composed of minute 'innumerable fascicles', each endowed with life and sensitivity.[39] He held that most tissues consisted of spherical monads. 'The sphere is, therefore, the most perfect form . . . The inorganic is angular, the organic spherical.'[40] This quotation exemplifies the ease with which notions of ideal form could be spun into metaphysical romances, which in fact occurred on a large scale. Nees von Esenbeck (1776–1858) published a *Textbook of Botany* (1820–1) interpreting the entire plant kingdom as one mighty leaf! The French zoologist Geoffroy St Hilaire (1772–1844) based his extensive system of comparative anatomy on the doctrine of an ideal fundamental type. Such writings reflect profound strivings for insight into the complexities of nature, coupled with disdain for empirical procedures. To these writers form was a transcendental quality not objectively based upon the organism. This approach to morphology was characteristic of the *Naturphilosophie*, a school of speculative unitary interpretations of phenomena guided by visionary ideas of nature.

Two examples may serve to show the limitations of the *Naturphilosophie* in the interpretation of form. The first naturalist to devote sustained systematic study to single-celled animals was Christian Gottfried Ehrenberg of Berlin (1795–1876). These minute forms of life have barely any permanent internal structure, although Ehrenberg was disposed to expect to find internal organs, prompted by the belief that all living beings had been created according to a single plan. The inclusions in the bodies of protozoa were sufficiently indistinct under microscopes then in use to afford considerable latitude in interpretation to the observer, and techniques of staining them to improve their visibility had not yet been developed. Ehrenberg thought that the liquid vacuoles that appear and move about within protozoa in the course of nutrition and liquid exchange were stomachs connected by an intestine. He believed he saw within protozoa perfect muscles, nerves, mouths, and veins. His name for them was 'stomach animals'. Some of the drawings for his monograph on infusoria in 1838 showed them with fully developed intestinal tracts.[41] The English anatomist Richard Owen (1804–92) subscribed to the same belief and portrayed vorticella as a whole series of stomachs arranged around a circular gut.[42] The belief in unity of plan conditioned the expectations of such observers and influenced their interpretation of much that they observed. The French microscopist Felix Dujardin (1801–62) argued strenuously against such interpretations, pointing out that organ systems could not exist below a certain threshold of size, for vessels would become too fine to take up water by capillary action. He argued that the form of the smallest animals was likely to be governed by the fluid properties of their substance and suggested inquiries into the types of symmetry characterizing their form. This was a highly fecund suggestion, leading toward the various groupings of protozoa based upon their symmetry properties by Haeckel and later systematists, as well as to the profound morphological insights of Otto Bütschli and D'Arcy Wentworth Thompson upon which much modern knowledge of these organisms is based.[43]

A second example was the tendency, stemming from Oken, to believe that all tissues were composed of globules. In the words of Joseph and Carl Wenzel the globule was 'the fundamental structure of all the solid parts without exception'.[44] Partly encouraged by spherical aberrations in the images of primitive microscopes a host of observers claimed to find globules everywhere in the fine structure of tissues. Sir Everard Home, an eminent English anatomist, wrote that all muscle and nerve structures were made up of tiny spherical units,[45] Such well-known anatomists as Johann Friedrich Meckel (1781–1833), Henri Milne-Edwards (1800–85), and Henri Dutrochet (1776–1847) fell under the spell.[46]

In 1838 and 1839 Matthias Schleiden (1804–81) and Theodore Schwann (1810–82) propounded a radically different explanation of tissue structure – that all living bodies were either cells (single-celled organisms) or composed of cells as a primary level of organization. This revolutionized morphology for, as Schleiden stated it, 'The form depends upon the manner in which the cells are combined together.'[47] The organism was shown to be composed of specialized units but they were only subordinate elements of the whole, cognate with simple monocellular organisms, not building blocks. The cell theory represented the objective application of the elements of the concept of organic form to the entire universe of living things. First, the fertilized germ cell governs the development of the whole and is antecedent to the differentiation of the parts. Second, the divisions among cells record the course of growth. Third, the cell imbibes nutrients and then divides, so that growth occurs by the assimilation of external elements into its own substance. Fourth, the achieved form is directed from within, although the means by which this occurred were by no means clear to the early proponents of the cell theory. Fifth, the cells are interdependent, functionally specialized and unable to exist apart from the organism except in tissue culture.[48]

It is noteworthy that Schleiden was completely unaware of the part played by aesthetic presuppositions in the development of his insights, partly because he was at pains to contrast his scientific manner of proceeding to the speculative character of the *Naturphilosophie.* He bitterly deplored the 'total want of any scientific principle' among idealistic morphologists. 'In that unbounded region every individual's imagination had naturally equal right.' Speculative writers on problems of form had 'stirred up together imagination and intellect, musing and thought, poetry and science, into a mixture as distasteful to the true poet as to the clear thinker'.[49] Schleiden was unaware that the criteria which he so successfully established as primary to the organization of life had been central to the literary aesthetics of Coleridge and Schlegel. If anyone had compounded imagination and intellect, poetry and science, it was Schleiden himself, to which we may perhaps obtain a clue from the title of his later popular work, *Poetry of the Vegetable World* (1853). The development of the cell theory consisted in the transformation of aesthetic presuppositions into scientific knowledge in a manner that strikingly vindicated Kant's assertion that the sense of beauty is an aid to the discovery of truth.

By ending the divorce between science and aesthetics, for so one may characterize Goethe's insistence upon the ideal character of beautiful forms, naturalists availed themselves of powerful aids to investigation, a conformity of the mind to its object. Such a stern rationalist as Thomas Henry Huxley (1825–95), the eloquent Victorian champion of rigor and strictness, made some of his most important discoveries as a result of an inherent aesthetic disposition to appreciate unities and symmetries in complex forms. He attacked Ehrenberg's notions of unity of plan as 'wonderful illustrations of what zoological and physiological reasoning should *not* be'.[50] He praised naturalists who were 'unable to see the propriety and advantage of introducing into science any ideal conception, . . . and who view with extreme aversion any attempt to introduce the phraseology and mode of thought of an obsolete and scholastic realism into biology'.[51] Yet it was through a penetrating visual analysis of the bewildering variety of siphonophore jellyfishes that Huxley established a unity of plan among coelenterates, on the strength of which he was made a Fellow of the Royal Society at the age of twenty-five and awarded its Royal Medal a year later.[52] It seems fitting, therefore, that this great biologist, distinguished by such powers of perception, would have contributed to the first issue of *Nature*, the celebrated journal of science, which he helped to found, his own translation of the fragment 'On Nature' (1780), the most impassioned lines Goethe ever wrote on his idea of nature. Huxley described the genesis of visual representations as 'That fashioning by Nature of a picture of herself, in the mind of man, which we call the progress of Science'.[53] How appropriate that he should observe, 'In travelling from one end to the other of the scale of life, we are taught one lesson, that living nature is not a mechanism but a poem; not a mere rough engine-house for the due keeping of pleasure and pain machines, but a palace whose foundations, indeed, are laid on the strictest and safest mechanical principles, but whose superstructure is a manifestation of the highest and noblest art.'[54]

In the poet's dream as related in the fifth book of *The Prelude* there appears a mysterious figure mounted on a camel, holding beneath one arm a stone and in the other hand 'a shell / Of a surpassing brightness'. The stone represents Euclid's *Elements*, 'purest bond / Of reason'. He held out the shell saying '"This . . . / Is something of more worth"';

> and at the word
> Stretched forth the shell, so beautiful in shape,
> In color so resplendent, with command
> That I should hold it to my ear. I did so,
> And heard that instant in an unknown tongue,
> Which yet I understood, articulate sounds,
> A loud prophetic blast of harmony;
> (Book V, lines 89–95)

The stone signifies the realm of inorganic nature to which simple principles of geometry apply. The shell stands for the world of living beings characterized by a higher geometry.

Shells were an eternal source of aesthetic delight. One of the most celebrated was the pearly nautilus, which was found between Fiji and the Philippines. The name had been used originally for the paper nautilus of the Mediterranean but was applied to the Pacific shells by Linnaeus. Not until 1832 was the organism that forms the shell described, in a celebrated memoir by Richard Owen consisting of careful description unaffected by notions of ideal form or unity of plan.[55] In 1821 the mathematician Sir John Leslie had suggested that the spiral was an 'organic' curve which resembled the 'elegant' subdivisions of the nautilus shell and was therefore suitable for ornamental use in architecture.[56] In a communication to the Royal Society in 1838 Canon Henry Moseley, professor of natural philosophy and astronomy in the University of Cambridge, reported the results of very careful measurements he had made from an illustration of the section of the pearly nautilus. He found that radii intercepted the spiral curve of the shell so that the width of the outer whorls was always three times that of the next, and that 'the curve is therefore an equiangular spiral' or logarithmic spiral.[57] This type of curve has a number of mathematically interrelated qualities, one being that the figure grows without changing its shape. No other mathematical curve shows this property. Moseley related this derivation to functions of nutrition, believing that he had found important evidence of the basic lawfulness and regularity of growth. The logarithmic spiral also defines the golden section, which had been considered since antiquity to be intimately related to considerations of ideal property and beauty. Here the very paridigm of beauty was discovered to exist objectively in the forms of organisms.

John Goodsir (1814–67), professor of anatomy at Edinburgh, attained distinction as a zoologist, parasitologist, cellular physiologist, and comparative anatomist. He helped to establish the importance of cells in nutrition and secretion. It is a mark of his importance that Rudolph Virchow dedicated his landmark work *Cellular Pathology* to him. Goodsir was much interested in Goethe's morphological writings and also in aesthetics, having helped to establish a discussion club on the subject in Edinburgh in 1851. One of the papers he read to this group was 'On the Natural Principles of Beauty', the same title as the treatise by his friend and colleague, D. R. Hay, in 1852, who is known also as the author of *Science of Beauty* (1856). Goodsir was fascinated by symmetry concepts and all manifestations of principles of form in animals. His biographer remarks that he possessed 'a vein of poesy mingled with a large aesthetic feeling that enhanced the beauty of form in his eyes, and rendered more patent the loveliness and adaptation in the mechanism and physiological operations in the varied structure of organisms.'[58]

Supposing the logarithmic spiral an emblem of the basic principle of organization in living nature, Goodsir figured it on the cover of the first part of his *Annals of Anatomy and Physiology* (1850) and expressed the hope that Moseley's demonstration of strict geometry in the nautilus might lead to Newtonian laws of organic growth.[59] One side of the obelisk raised to his memory in the Dean Cemetery in Edinburgh bears an incised spiral curve, symbolizing the law of vital force he had hoped to discover and also marking him a devout adherent of the idea of organic form. It is further evidence of the aesthetic appeal of the demonstration of the proportions of the nautilus shell that one of the most famous of American poems, by Oliver Wendell Holmes, describes how the creature progresses from chamber to chamber as the shell develops, leaving its 'low-vaulted past' to build 'more stately mansions':

> Year after year held the silent toil
> That spread his lustrous coil;
> Still, as the spiral grew,

> He left the past year's dwelling for the new,
> Stole with soft step its shining archway through,
> Built up its idle door,
> Stretched in his last-found home, and knew the old no more.

In a related development that cannot but impress one as unexpected and remarkable, the botanists K. F. Shimper (in 1835), A. Braun (in 1831 and 1835) and the brothers A. and L. Bravais (in 1837) showed that the spiral lines that could be drawn through leaf attachments (in the fifth of Bonnet's classes of leaf arrangement) showed mathematical properties akin to those of the logarithmic curve. Such spirals might be written as fractions in which the denominator indicates the number of leaves that appear before another leaf is inserted in the same vertical plane as the first, while the numerator describes the number of rotations around the axis by which this takes place. The fractions found generally to occur in nature may be arranged in the series $\frac{1}{2}$, $\frac{1}{3}$, $\frac{2}{5}$, $\frac{3}{8}$, $\frac{5}{13}$, $\frac{8}{21}$, $\frac{13}{34}$, $\frac{21}{55}$, which is the same as that which results from the expansion into a continuous fraction of the irrational number $\frac{1}{2}(\sqrt{5}-1)$. This number is the ratio known as the golden section. The assemblage of scales in pine cones or the arrangement of the fruits of the pineapple were found to consist of overlapping spirals that conform to the series. The floret and seed arrays of the sunflower manifest 34 short spirals and 55 long ones, thus exemplifying the same numerical properties.[60] Thus plant forms upon analysis yielded an irrational number, providing an absolute distinction from the forms of crystals, whose faces, in accord with Haüy's law, must be reducible to rational numbers. The golden section took on almost mystical significance as the principle at once of life and of beauty. And the spiral continues to yield significant intuitions of natural form, as in the celebrated double helix which represents the structure of the genetic material. James Watson has described his search for this structure as being greatly reinforced by seeing helical staircases during a visit to Oxford. He then pored over long series of electron photomicrographs of muscle and collagen for instances of helical symmetry, as though to train his eye to analyse the X-ray photographs of tobacco mosaic virus where he first recognized the helical form of RNA. His success in showing how base pairs could be assembled in double-helix form demonstrated a profoundly important empirical reality while at the same time confirming a presupposition of helical symmetry that 'was too pretty not to be true'.[61]

In a very important paper the Harvard physicist Gerald Holton has reappraised the origins of Einstein's special theory of relativity, frequently supposed to have been prompted by A. A. Michelson's demonstration that the passage of light through the atmosphere was unaffected by a supposed drift of ether at the earth's surface. Holton cites a number of statements by Einstein in which he denied that it had been the results of Michelson's experiment that had prompted him to re-examine electrodynamic theory and concludes instead that the 'primary impetus for Einstein was the essential requirement of finding symmetry and universality in the operations of nature'.[62] The determination of the ultrastructure of organic tissues and the configuration of molecules involves biologists in the study of 'conformation', as John C. Kendrew calls it, which offers ample scope to modes of visual analysis. 'To define what is new about it is actually quite difficult to do in any neat phrase, but in the last resort it is a question of geometry and symmetry.'[63] The Nobel prize-winning anatomist Santiago Ramón y Cajal (1852–1934) acknowledged a 'congenital inclination to economy of mental effort and the almost irresistible propensity to regard as true what satisfies our aesthetic sensibility by appearing in agreeable and harmonious architectural forms. As always, reason is silent before beauty.'[64] Ernst Haeckel, who called his microscope 'the dear companion of my life', acknowledged within himself 'a real sensuous element, which permits me to conceive and retain thoughts and facts, and to imprint them on my mind much more strongly than when they are merely represented dryly and nakedly in words.'[65] The validity of applications of symmetry principles to the analysis of natural form allows the biologist to draw upon the faculties of visual imagination and the aesthetic sense of harmony in evaluating concepts of structure.[66]

I do not suggest that biologists have in any way compromised the laws of scientific evidence in their treatment of form. In his celebrated summary treatise on morphology, *On Growth and Form* (1917), Sir D'Arcy Wentworth Thompson (1860–1948) interpreted the major phenomena of form, from the fluid shapes of membranes to the vertebrate skeleton, with unexcelled mathematical precision and close attention to physical mechanisms. 'The symmetry which the organism displays seems identical with that symmetry of forces which results from the play and interplay of surface-tensions in the whole system.'[67] Yet it is worth recalling that Thompson was a superb prose stylist and a classical scholar whose translation of Aristotle's *Historia Animalium* may possibly never be surpassed.[68] The elegance and scope of his analysis has made *On Growth and Form* a source of inspiration for countless artists. The protozoologist Clifford Dobell called it 'a work of art no less than of science'.[69] Discussions by such outstanding scientists as Joseph Needham, C. H. Waddington, Paul Weiss, and Edmund Sinnott leave no room for impressions that the definition of organic form lacks rigor and demonstrate exhaustively and with a wealth of examples the essential difference of the form of organisms from the properties of non-living matter.[70]

Writers on scientific method usually emphasize the philosophical rigor by which explanations must be related to the data of observation and experiment. As I have sought to trace the sources of influence upon the history of the concept of organic form I have become increasingly sensible of the limitations inherent in the tendency to study science as an autonomous process developing as a strict logical sequence of chaste hypotheses subjected to confirmation or rejection. Increasingly I have felt the need to refer to the history of imagination and perception – processes pervaded by other cultural influences.[71] The history of morphology seems to bear out the thesis recently applied to the history of physics by A. M. Taylor, that the 'artistically creative imagination' in Max Planck's phrase has played a primary role in the development of science.[72] Scientists have been able to deal successfully with the processes of growth only through the exercise of a poetic faculty responding sensitively to processes peculiar to the world of organisms. Efforts to extend notions drawn from the study of less complex phenomena in chemistry and physics to the realm of living form have repeatedly led to error and inadequate interpretations. So there was gradually built up in the scientific portrayal of life forms an aesthetic endowment. In a book-length study, *The Art of Organic Forms* (1968), I have tried to show how this occurred and how painters and sculptors, most notably Klee and the Surrealists, drew upon that endowment in works which recreate the world of nature in strict compliance with the formal precepts of the idea of organic form.[73] Compositions by Gorky, Matta, Arp, and many others evince curvilinear motifs which are images of growth, interdependence of parts, and differentiation directed from within. Their compositions reinterpret the basic principles of organic form. They do not, with some exceptions, represent organisms directly. Through them a rigorous if somewhat protean scientific concept has entered the culture at large.

So pervasive and many-sided a cultural development deserves a wider interpretation than can be found for it in the history of aesthetics or the history of science alone, and to that purpose this volume is dedicated. We seem to be witnessing a widening undercurrent flowing against the mechanistic and technologically oriented surge which has borne western society along for at least two centuries. The rise of the idea of organic form attends and has partly helped to bring about a change from a physical to a biological basis in our conception of society, bespeaking a harmony with the world that gave us birth and a heightened sense of affinity with living nature. Lewis Mumford has documented the extent of the sway exercised over society by mechanistic concepts, entailing vast sacrifices of amenity and freedom in order to concentrate on unprincipled growth of power and wealth.[74] He foresees a reordering of the human enterprise around organic modes of response and integration. How vast and consequential has become the challenge to the mechanical world view since Coleridge and Wordsworth professed their belief in an order peculiar to the realm of life! One of the central images figuring mightily in Thomas Mann's *Doctor Faustus* (1947) is man's arrogant attempt to imitate organic form through physical or chemical means: ice crystals, drops of oil, and chemical growths whereby Adrian Leverkühn's father sought to show that any organic form could be duplicated by mechanical means. This was the impious laboratory in which were conceived extravagant impulses to outstrip nature in creativity. Dare man tamper with the primal forces of nature or ignore the special properties of life? Dare man interrupt nature's dream of herself with music of his own? What would transpire if his own melody were to drown out the rhythms which are so deeply imprinted within him by his biological and social inheritance? This is the Faustian endeavor and the curse of this age, to which the dying composer confesses: 'instead of shrewdly concerning themselves with what is needful upon earth that it may be better there, and discreetly doing it, that among men such order shall be established that again for the beautiful work living soil and true harmony be prepared, man playeth the truant and breaketh out in hellish drunkenness; so giveth he his soul thereto and cometh among the carrion.'[75]

Notes

1 Lynn White, Jnr, 'Natural science and naturalistic art in the Middle Ages', *American Historical Review*, LII (1947), p. 421.
2 V. P. Zubov, *Leonardo da Vinci*, trans. David H. Kraus, Cambridge, Mass.: Harvard University Press, 1968, p. 57.
3 Panofsky, *The Life and Art of Albrecht Dürer*, Princeton University Press, 1955, p. 18. His discussion of 'Melencolia I' is of great interest in relation to criticisms of Cartesianism, pp. 156–72.
4 P. C. Ritterbush, 'Art and science as influences on the early development of natural history collections', *Proceedings of the Biological Society of Washington*, LXXXII (1969), pp. 561–78 and works cited.
5 P. C. Ritterbush, *Overtures to Biology; The Speculations of Eighteenth-Century Naturalists*, New Haven: Yale University Press, 1964, pp. 109–22 for works cited.
6 [Linnaeus], 'Metamorphosis Plantarum' [1755], Peter Bremer, *Amoenitates Academicae* (Leyden edition), IV (1759), pp. 368–86.
7 *Rêveries du promeneur solitaire* [1782]; *Confessions* . . ., London: J. Bew, 1783, II, pp. 215–16. See also his letter of 21 September 1771 to Linnaeus, in James Edward Smith, ed., *A Selection of the Correspondence of Linnaeus*, London: Longman, Hurst, Rees, Orme, and Brown, 1821, II, p. 553.
8 Letter to Thomas Percival, 8 August 1785. *The Works . . . of Thomas Percival*, London: J. Johnson, 1807, III, p. cv. On plant sensitivity see *Overtures to Biology* (cited in n. 5), pp. 141–74 and P. C. Ritterbush, 'John Lindsay and the sensitive plant', *Annals of Science*, XVIII (1962), pp. 233–54.
9 Darwin, *The Loves of the Plants* (1789), Canto I. In this note he originated the theory of mimicry, noting that many flowers resemble insects. He believed this was a mechanism for protecting nectar by discouraging insects from landing on the flowers, which was not then

known to be necessary for fertilization. The error does not detract from this important recognition that resemblance to other organisms may have biological functions. At another point, Additional Note 39 to *The Economy of Vegetation* (1791), Darwin conjectured that insects may have originally evolved as flowers that had somehow worked loose from their parent plant, a suggestion repeated by Coleridge as a 'Darwinian flight' in his essay, 'On the definitions of life hitherto received. Hints toward a more comprehensive theory' [1830] in *Miscellanies, Aesthetic and Literary*, London: George Bell, 1885, p. 413. It is remarkable to find Coleridge sharing in the sort of speculation that he himself condemned as 'Darwinising with a vengeance', 'Notes on Stillingfleet', *Athenaeum*, no. 2474 (27 March 1875), pp. 422–3.

10 *Erasmus Darwin*, London: Macmillan, 1963. Reviewed by P. C. Ritterbush in *Science*, CXLIII (6 March 1964), pp. 1024–5.

11 See *The Economy of Vegetation* (1791), 'Apology' and numerous notes, including Additional Note XXII; also *Overtures to Biology* (cited above in n. 5), pp. 172–3 for works cited.

12 Cf. C. H. Waddington, *Behind Appearance. A Study of the Relations Between Painting and the Natural Sciences in This Century*, Cambridge, Mass.: M.I.T. Press, 1970, esp. pp. 147–82 and the quotation from Naum Gabo on p. 46. Reviewed by P. C. Ritterbush in *Science*, CLXIX (21 August 1970), pp. 751–2.

13 Linnaeus, *Philosophia Botanica*, Stockholm: Godfrey Kiesewetter, 1751, p. 48.

14 Charles Bonnet, *Recherches sur l'usage des feuilles dans les plantes*, Göttingen and Leyden: Elie Luzac, Fils, 1754, p. 164 and Plate 20.

15 John Beaumont, 'Two letters written by Mr. John Beaumont Junior of Stony-Euston in Somersetshire, concerning rock-plants and their growth', *Philosophical Transactions*, XI (1667), p. 724.

16 N. Grew, *The Anatomy of Plants*, London: for the author, 1682, p. 159.

17 See John G. Burke, *Origins of the Science of Crystals*, Berkeley: University of California Press, 1966. Also J. D. Bernal, 'Symmetry of the genesis of form', *Journal of Molecular Biology*, XXIV (1967), pp. 379–90.

18 Young, *Conjectures on Original Composition*, ed. Edith J. Morley, Manchester University Press, 1918, p. 7.

19 M. H. Abrams, *The Mirror and the Lamp: Romantic Theory and the Critical Tradition*, New York: Oxford University Press, 1953. Cf. the use of the plant as an emblem of the human sensibility by Carl Gustav Carus, *Natur und Idee*, Leipzig: Wilhelm Braumüller, 1861, following p. 459.

20 Ibid., p. 119.

21 For an example drawn from the sciences, see Sir Humphry Davy, lecture of 1811, in John Davy, ed., *The Collected Works of Sir Humphry Davy* . . . , London: Smith, Elder, 1840, VIII, pp.308, 317.

22 H. E. Gerlach and O. Herrmann, *Goethe erzählt sein Leben*, Frankfurt: Fischer Bücherei, 1956, pp. 218–19.

23 Ibid., p. 246.

24 Goethe, Review of J. B. Wilbrand and F. A. von Ritgen, *Gemälde der organischen Natur in ihrer Verbreitung auf der Erde*, Giessen: C. G. Müller, 1821, in Bertha Mueller, ed., *Goethe's Botanical Writings*, Honolulu: University of Hawaii Press, 1952, pp. 120–1.

25 Goethe, 'Einfache Nachahmung der Natur, Manier, Stil' [1788], *Goethes Werke*, Stuttgart: J. G. Cotta, 1868, XXVII, p. 28.

26 C. S. Sherrington, *Goethe on Nature and on Science*, Cambridge University Press, 1942, p. 27.

27 For a translation and interpretation of these poems see Elizabeth Sewell, *The Orphic Voice: Poetry and Natural History*, New Haven: Yale University Press, 1960, pp. 269–75.

28 Cf. William Morton Wheeler, 'The ant-colony as an organism' [1910], reprinted in *Essays in Philosophical Biology*, Cambridge: Harvard University Press, 1939; and Edmond Perrier, *Les Colonies animales et la formation des organismes*, 2nd edition, Paris: Masson et cie., 1898.

29 Jacques Barzun and Ralph H. Bowen, *Diderot; Rameau's Nephew and Other Works*, Garden City, New York: Doubleday Anchor Books, 1956, p. 119.

30 W. K. Wimsatt, Jr, *Literary Criticism; A Short History*, New York: Alfred A. Knopf, 1957, p. 371. Cf. Lewis Mumford on the evolutionary roots of man's visual affinities for the world of plants, *The Myth of the Machine; The Pentagon of Power*, New York: Harcourt Brace Jovanovich, Inc., 1970, pp. 378–84.

31 E. Cassirer, 'Goethe and the Kantian philosopher', in *Rousseau, Kant, Goethe: Two Essays*, trans. by James Gutmann *et al.*, Princeton University Press, 1970, pp. 61–98. For the place of Kant's ideas in the history of concepts of design in nature see Clarence Glacken, *Traces on the Rhodian Shore; Nature and Culture in Western Thought from Ancient Times to the End of the Eighteenth Century*, Berkeley: University of California Press, 1967, pp. 530–7.

32 I. A. Richards, *Coleridge on Imagination*, London: Kegan Paul, Trench, Trubner, 1934; also Robert D. Hume, 'Kant and Coleridge on imagination', *Journal of Aesthetics and Art Criticism*, XXVIII (1970), pp. 485–96.

33 In T. N. Raysor, ed., *Coleridge's Shakespearean Criticism*, London: Constable, 1930, I, p. 224; see also pp. 4, 5. Also see Joseph Needham, 'Coleridge as a philosophical biologist', *Science Progress*, XX (1926), pp. 692–702.

34 In T. Ashe, ed., *The Table Talk and Omniana of Samuel Taylor Coleridge*, London: G. Bell, 1923, p. 145 [18 December 1831].

35 In Kathleen Coburn, ed., *The Philosophical Lectures of Samuel Taylor Coleridge Hitherto Unpublished*, New York: Philosophical Library, 1949, p. 196 [18 January 1819].

36 'Monologues by the Late Samuel Taylor Coleridge, Esq. No. I: Life', *Fraser's Magazine for Town and Country*, XII (1835), p. 495.

37 Alice D. Snyder, ed., *S. T. Coleridge's Treatise on Method*, London: Constable, 1934, p. 7.
38 In Jakob Minor, ed., *Friedrich Schlegel: Seine prosaischen Jugendschriften*, Vienna: C. Konegen, 1906, II.
39 L. Oken, *Elements of Physiophilosophy* [1809–11], trans. Alfred Tulk, London: Ray Society, 1847, p. 190.
40 Ibid., p. 29.
41 C. G. Ehrenberg, *Die Infusionsthierchen als vollkommene Organismen: Ein Blick in das tiefere organischen Leben der Natur*, Leipzig: Leopold Voss, 1838.
42 R. Owen, *Lectures on the Comparative Anatomy and Physiology of the Invertebrate Animals*, London: Longman, Brown, Green, & Longman, 1843, pp. 25–6.
43 Felix Dujardin, *Histoire naturelle des zoophytes: Infusoires*, Paris: Librairies Encyclopédique de Roret, 1841, p. 24 ff.; E. H. Haeckel, *Generelle Morphologie der Organismen: Allgemeine Grundzüge der organischen Formen Wissenschaft*, Berlin: Georg Reimer, 1866, two vols, presents three primary groups which generally resemble those recognized by the outstanding modern authority Libbie H. Hyman, *The Invertebrates: Protozoa through Ctenophora*, New York: McGraw-Hill, 1940, pp. 18–21 and Figure 4.
44 J. and C. Wenzel, *De Penitori Structura Cerebri Hominis et Brutorum*, Tübingen: Cottam, 1821, p. 27 ff.
45 E. Home, 'In the changes the blood undergoes in the act of coagulation', *Philosophical Transactions*, CVIII (1818), pp. 172–98 and idem., CXI (1821), pp. 26–46.
46 J. F. Meckel, *System der vergleichenden Anatomie*, Halle: Rengerschen Buchhandlung, 1821, I, pp. 38–43, and *Manual of General, Descriptive, and Pathological Anatomy* [1816–1820], trans., A. Sidney Doane et al., Philadelphia: Carey and Lea, 1832, I, pp. 22–5. H. Milne-Edwards, 'Mémoire sur la structure élémentaire des principaux tissus organiques des animaux', *Archives générales de médicine . . .*, 1st year, III (September 1823), pp. 165–84. H. Dutrochet, *Recherches anatomiques et physiologiques sur la structure intime des animaux et des végétaux, et sur leur motilité*, Paris: J.-B. Ballière, 1824, p. 215.
47 M. J. Schleiden, *Principles of Scientific Botany . . .* [1842], trans. Edwin Lankester, London: Longman, Brown, Green, & Longman, 1849, p. 135.
48 See John R. Baker, 'The cell theory: a restatement, history, and critique', *Quarterly Journal of Microscopical Science*, LXXXIX (1948), pp. 103–25, XC (1949), pp. 87–108 and p. 331, XCIII (1952), pp. 157–89, XCIV (1953), pp. 407–40, and XCVI (1955), pp. 449–81.
49 Schleiden, op. cit., p. 313.
50 T. H. Huxley, 'Zoological notes and observations . . .' [1851], in Michael Foster and E. Ray Lankester, eds, *The Scientific Memoirs of Thomas Henry Huxley*, London: Macmillan, 1898–1902, I, p. 89.

51 T. H. Huxley, 'On the theory of the vertebrate skull' [1858], *idem.*, pp. 584–5.
52 T. H. Huxley, 'On the anatomy and the affinities of the family of the Medusæ' [1849], idem., pp. 9–32.
53 T. H. Huxley, in *Nature*, I, no. 1 (4 November 1869), p. 10.
54 T. H. Huxley, 'On natural history, as knowledge, discipline and power' [1856], loc. cit., p. 311.
55 R. Owen, *Memoir on the Pearly Nautilus*, London: W. Wood, 1832.
56 J. Leslie, *Geometrical Analysis and Geometry of Curve Lines*, Edinburgh: W. & C. Tait, 1821, p. 438.
57 H. Moseley, 'On the geometrical form of turbinated and discoid shells', *Philosophical Transactions*, CXXVIII (1838), p. 361.
58 Henry Lonsdale, 'Biographical memoir', in *The Anatomical Memoirs of John Goodsir, F.R.S.*, Edinburgh: Adam & Charles Black, 1868, I, pp. 126–7. See Jay Hambridge, *Dynamic Symmetry: The Greek Vase*, New Haven: Yale University Press, 1920.
59 J. Goodsir, 'On the employment of mathematical modes of investigation in the determination of organic forms', ibid., II, pp. 205–19.
60 Alexander Braun, 'Vergleichende Untersuchung über die Ordnung der Schuppen an den Tannenzapfen als Einleitung zur Untersuchung der Blattstellung', *Nova Acta Physico-Medica Academiae Caesareae Leopoldino-Carolinae*, XVII (1831), pp. 195–402. Also A. H. Church, *Types of Floral Mechanism* . . . Part 1, Oxford: Clarendon Press, 1908 and *On the Interpretation of Phenomena of Phyllotaxis*, 'Botanical Memoirs', Oxford University Press, 1920. For discussion of related instances of the symmetry principles of organisms as differing from those of inanimate matter see Martin Gardner, *The Ambidextrous Universe*, New York: Basic Books, 1964, pp. 105–32; Helena Curtis, *The Viruses*, New York: Natural History Press, 1965, pp. 163–81; and F. M. Jaeger, *Lectures on the Principle of Symmetry and Its Applications in All Natural Sciences*, 2nd edition, Amsterdam: Elsevier, 1920, pp. 160–70.
61 James Watson, 'The double helix: the discovery of the structure of DNA', Part 2, *The Atlantic Monthly*, CCXXI (1968), p. 114.
62 G. Holton, 'Einstein, Michelson, and the "Crucial" Experiment', *ISIS*, LX (1969), p. 192. Holton calls for the development of 'a field that can fairly be called the aesthetics of science', p. 177.
63 J. C. Kendrew, 'Information and conformation in biology', in Alexander Rich and Norman Davidson, eds, *Structural Chemistry and Molecular Biology*, San Francisco and London: W. H. Freeman, 1968, p. 196.
64 Ramón y Cajal, *Recollections of My Life*, trans. E. Horne Craigie, American Philosophical Society, *Memoirs*, VIII, Part 2 (1937), p. 303.
65 E. H. Haeckel, letter of 25 December 1852, *The Story of the Development of a Youth*, New York: Harper, 1923, p.56.

66 On systematic biology see C. F. A. Pantin on 'aesthetic recognition' in 'The recognition of species', *Science Progress*, XLII (1954), p. 587. On symmetry concepts in the elucidation of higher taxa see Michael J. Greenberg, 'Ancestors, embryos, and symmetry', *Systematic Zoology*, VIII (1959), pp. 212–21. Konrad Lorenz has written that no one could bring to bear on animals the powers of concentration required for the analysis of behavior 'unless his eyes were bound to the object of his observation in that spellbound gaze which is not motivated by any conscious effort to gain knowledge, but that mysterious charm that the beauty of living creatures works on some of us'. 'Physiological mechanisms in animal behavior', Symposia of the Society for Experimental Biology, no. 4, Cambridge, 1950, p. 235. On embryology see Ross Granville Harrison, 'Relations of symmetry in the developing embryo', *Organisation and Development of the Embryo*, New Haven: Yale University Press, 1969, pp. 166–214 and Bernal, cited above in n. 17. On biochemistry see Patrick D. Ritchie, 'Recent views on asymmetric synthesis and related processes', in F. F. Nord, ed., *Advances in Enzymology*, New York: Interscience Publishers, 1947, pp. 65–110 and George Wald, 'The origins of life', *Proceedings of the National Academy of Sciences*, LII (1964), pp. 595–611. On virus forms see F. H. C. Crick and J. D. Watson, 'Structure of small viruses', *Nature*, CLXXVII (1956), pp. 473–5. Useful summaries are those of Jacques Monod, 'On symmetry and function in biological systems' in Arne Engström and Bror Stranberg, eds, *Symmetry and Function of Biological Systems at the Macromolecular Level; Proceedings of the Eleventh Nobel Symposium*, New York: Wiley Interscience Division, 1969, pp. 15–27 and Hermann Weyl, *Symmetry*, Princeton University Press, 1952.

67 D'A. W. Thompson, *On Growth and Form*, 2nd edition, New York: Macmillan, 1943, p. 345. See also his remarks on 'the harmony of the world' and 'mathematical beauty', pp. 1096–7. See also C. M. Wardlaw, 'Mathematical relationships', in *Morphogenesis in Plants*, London: Methuen, 1952, pp. 73–90.

68 Cf. D'A. W. Thompson, *Science and the Classics*, St Andrews University Publication no. 44, Oxford University Press, 1940.

69 C. Dobell, 'D'Arcy Wentworth Thompson', *Obituary Notices of the Royal Society of London*, n. 18 (1949), p. 612.

70 P. Weiss, 'Beauty and the Beast; Life and the Rule of Order', *Scientific Monthly*, LXXXI (1955), pp. 286–99; Conrad H. Waddington, *The Nature of Life*, London: Allen & Unwin, 1961; J. Needham, *Order and Life*, New Haven: Yale University Press, 1936; reprinted by the M.I.T. Press, 1968; and E. W. Sinnott, *The Problem of Organic Form*, New Haven: Yale University Press, 1963.

71 Cf. P. C. Ritterbush, 'The shape of things seen: the interpretation of form in biology', *Leonardo*, III (1970), pp. 305–17.

72 A. M. Taylor, *Imagination and the Growth of Science*, London: John Murray,

1966. See also the early and important study by Martin Johnson, *Art and Scientific Thought; Historical Studies towards a Modern Revision of Their Antagonism, with a Foreword by Walter de la Mare*, London: Faber & Faber, 1944.

73 P. C. Ritterbush, *The Art of Organic Forms*, Washington: Smithsonian Institution Press, 1968; distributed by George Braziller, Inc. of New York and David & Charles, Newton Abbot.

74 See Lewis Mumford, 'The failure of mechanomorphism', *The Myth of the Machine*, 1970, cited above in n. 30, pp. 95–8. See Michael Polanyi, 'Life transcending physics and chemistry', *Chemical and Engineering News* (21 August 1967), pp. 54–66. Also Georges Canguilhem, *La Connaissance de la vie*, 2nd edition, Paris: Librairie Philosophique J. Vrin, 1965, pp. 101–27 and 'The role of analogies and models in biological discovery', in Alistair Crombie, ed., *Scientific Change*, New York: Basic Books, 1963, pp. 507–20. Also see René Dubos, *Reason Awake: Science for Man*, New York: Columbia University Press, 1969, p. 49 on 'subconscious ways' in which 'the dreams of mankind may still be influencing the orientation of scientific effort'.

75 T. Mann, *Doctor Faustus, The Life of the German Composer Adrian Leverkühn as Told by a Friend*, trans. H. T. Lowe-Porter, London: Penguin Books, 1968, p. 479 and ch. 3. For the antecedents to the foam drops and chemical imitations of organic form see *The Art of Organic Forms*, 1968, cited above in n. 73, pp. 63–76.

William K. Wimsatt

Organic form: some questions about a metaphor

The metaphor is ancient, and most of the questions have been asked before, many times. That I should write a paper of about twenty pages pretending to say anything worth while upon such a classic theme requires a courage that I derive largely from the support of my two collaborators – one a historian who has made himself an expert guide in that tropical rain-forest of eighteenth-century romantic nature philosophy and early scientific 'representations' of living form, in which some of our most cherished modern notions about both science and poetry had their seed-bed and first growth; and the other an idealist aesthetician and critic of imposing authority. I mean that it is not up to me to expound very much history, either of science or of aesthetics. I can take history as an object that is before us, almost palpably, upon the table, and I can choose my own exhibits. One of the first that I call attention to I take from a play by Moliere, *Don Juan, or the Feast with the Statue.* The agile valet Sganarelle, in one of his several running debates with his atheistic master, bursts into teleology:

> Can you perceive all the contrivances of which the human mechanism is composed without wondering at the way the parts are fitted with one another? These nerves, these bones, these veins, these arteries, these . . . this lung, this heart, this liver, and all the other organs . . . My argument is that there is something mysterious in man which, whatever you may say, none of the philosophers can explain. Is it not wonderful that I exist and that I have something in my head which thinks a hundred different things in a moment and does what it wills with my body? I wish to clap my hands, to raise my arms, to lift my eyes to heaven, to bow my head, to move my feet, to go to the right, to the left, forward, backward, to turn . . .
> (He falls down whilst turning.)
> Don Juan. Good, so your argument has broken its nose.
>
> *Don Juan*, III, i

A centuries-old Aristotelian and scholastic commonplace – 'What a piece of work is a man!' – is good enough material here for a hamstrung sprint of servile boldness, a bumbling theology. Larger versions of organicism appear in the same era with a kind of cool solemnity – as in the cosmic unity, the social and ethical harmonies, which critics nowadays celebrate and annotate from the whole Western tradition in the *Essay on Man* of Pope or his *Windsor Forest*.[1]

A quicker pulse, a new accent of excitement, marks, I believe, my second exhibit – if not a subtlety and coolness equal to that of either Molière or Pope.[2]

> What Beaux and Beauties crowd the gaudy groves,
> And woo and win their vegetable Loves . . .
> The love-sick Violet, and the Primrose pale,
> Bow their sweet heads, and whisper to the gale;
> With secret sighs the Virgin Lily droops,
> And jealous Cowslips hang their tawny cups.
> How the young Rose in beauty's damask pride
> Drinks the warm blushes of his bashful bride.
>
> BOTANIC MUSE! who in this latter age
> Led by your airy hand the Swedish sage,
> Bade his keen eye your secret haunts explore
> On dewy dell, high wood, and winding shore;
> Say on each leaf how tiny Graces dwell;
> How laugh the Pleasures in a blossom's bell.
> How insect Loves arise on cobweb wings,
> Aim their light shafts, and point their little stings.

The author of these heroic couplets, the grandfather of Charles Darwin, tells us in a footnote that he is drawing upon 'Linneus, the celebrated Swedish naturalist', and in an advertisement to the volume, he says that 'The general design . . . is to inlist Imagination under the banner of Science; and to lead her votaries from the looser analogies, which dress out the imagery of poetry, to the stricter ones, which form the ratiocination of philosophy.' The enlistment of the imagination and the looser analogies were a more successful part of the program than the stricter analogies. 'Eighteenth-century naturalists', Dr Ritterbush instructs us, 'denied or overlooked every distinction between plants and animals that they might have been expected to consider.'[3] The frontispiece of a German edition of Alexander von Humboldt's *Journey to the Equatorial Regions of the New World* (1807), transplanted as frontispiece of Ritterbush's *Overtures to Biology* (1964), shows us an Apollonian figure, the spirit of poetry no doubt, unveiling an Asiatic Artemisian statue, appropriate emblem of the mysterious fecundity of nature; the title of Goethe's *Elegie* on the growth of plants (*Die Metamorphose der Pflanzen*, 1799) is inscribed on a tablet lying at the feet of the multimammiferous goddess. Such celebrations were surely not harmful for the development of Romantic nature poetry. At the same time, a degree of confusion in poetic theory may have been a concomitant, even an inspiration, of the poetry. It is perhaps significant that the English Romantic poet who described nature (that is, English landscape) the most often and the most lovingly – Wordsworth of course – had in his theoretical essays little to say about organic form. Yet, by and large, the prevalence of nature, especially landscape and botanical nature, in English poetry during about two centuries, does suggest a kind of latent equation in the poetic mind: themes or images of organic life in poetry confer upon that poetry the poetic life of organic form. Coleridge, the most important translator of German organic idealism to the English scene, could speak, or seem to speak, both for and against that equation – and in the same essay. 'If the artist copies the mere nature, the *natura naturata*, what idle rivalry.'[4] On the other hand, it did seem to him that the visible image of nature was in a special way 'fitted to the limits of the human mind'. Natural forms, in a very natural way, yielded moral reflections; nature was thought, and thought nature. That was the 'mystery of genius in the Fine Arts'.[5] This might be illustrated by many passages of description and simile in the poetry of both Coleridge and Wordsworth or of Shelley. But, as Bernard Blackstone has pointed out, there is no English Romantic poet better than Keats for showing us the genial swell of organic forms, the unfolded buds, the ripening fruit, the loaded blessing of the vines, the swollen gourd, the plumped hazel shells, and for making such images symbolize a transcendent experience of beauty – even like that of a Grecian Urn. The first volume of Erasmus Darwin's *Botanic Garden* opened with engravings of flowers, and the second volume concluded with engravings by William Blake of the Portland vase.[6]

What then of organic form in visual art? The notebooks of Goethe and a crowd of other Nature Philosophers and scientists and their treatises and textbooks are lavish with both pictures and scientific 'representations' or modules of organic life. But these no doubt serve little enough any aesthetic purpose. The 'meticulously veined leaves' painted under the Pre-Raphaelite hand lens of Millais, or Ruskin sprawled out drawing a square foot of meadow grass on a mossy bank,[7] are more valid indexes to our theme, and no less the vegetable curves of Art Nouveau at the end of the century – Hector Guimard's sinuously framing metallic green tendrils and leaves for Paris Métro station portals, for instance, one of which we can see today in New York, in the sculpture garden of the Museum of Modern Art. Let a recent historian of that era in art have the last word about this: 'Up from the sidewalk there sprang a profusion of interlacing metal, bouquets of aquatic plants, luminous tulips, gorged with the rich disturbing sap of Paris, its cellars, and subsoil.'[8] The date was about 1900 – at about the same date the Parisian aesthete Gustave Geffroy wrote the following encomium upon a flower: '*A flower* . . . Free and growing out of the earth, or captive in a vase, it presents an artist with the perfect example of the universal creative force – in it he may find form, colour and even expression, a mysterious expression composed of stillness, silence, and the fugitive beauty of things which are born only to die in the same moment.'[9]

And this brings us to the art exhibition put on by Dr Ritterbush at the Museum of Natural History in Washington D.C., during June and July of 1968 – *The Art of Organic Forms*: cells, globules, curves, filaments, membranes, tentacles, emulsions, gelatins, pulsing fluids, capillary action, liquid diffusions, amoeboid shapes, nascent protoplasmic entities[10] – all this and more concentrated in seventy-two paintings, graphics, and sculptures of our own century: Odilon Redon, for instance, *Au fond de la Mer*, c. 1905; Paul Klee, *Male and Female Plants*, 1921; Wasily Kandinsky, *Capricious Forms No. 643*, 1937; Max Ernst, *Prenez garde au microbe de l'amour*, 1949; Pavel Tchelichew, *Itinerary of Light*, 1953. Matta's *Le Vertigo d'Erôs*, 1944, at the Museum of Modern Art, supplies the mysteriously greenish ektachrome frontispiece of the *Catalogue* of this remarkable exhibition. The thesis of many of these artists and of their critics, expressed in essays and catalogue notes, is that such submicroscopic life forms symbolize the secret life of the creative human spirit.

We have been skirting a sophism: namely, the notion that the representation of biological forms in a work of verbal or visual art implies something about the presence of organic or artistic form in that work.[11] An idealist historian and philosopher such as Professor Orsini will not wish to linger long in discussion of that issue. Nevertheless, its recurrent presence, even as a hint or as a half-committed fallacy, throughout the now long stretch of the modern organistic era, justifies its being noted and put aside with some deliberacy at the start of a discussion that aims at the center of the critical question of organicism. The issue is perhaps more easily defined in literary art, perhaps more nicely subject to confusion in visual art. Consider the following paradox. If the picture is overt enough, say a Currier and Ives print of water-melon vines, trumpet flowers, and humming birds, it presents organic forms, but I think none of us will be likely to argue that it thereby *has* high artistic form. Move the picture, however, through several shades of abstractionism, say through Art Nouveau sinuosities to pure or supposedly pure non-referential curves, and then to the golden-section compositional style of Pieter Mondrian or the 'illusionary modulations', *Despite Straight Lines*, of Josef Albers.[12] We reach a stage where so far as the picture has content, it is a geometric content. But this too is a geometric form, for geometry is all form. Form and representational content coincide perfectly. So in a sense this must be artistic form, and hence, by idealistic definition, it must be organic form. Yet it is not life form, but rather crystal form, as Dr Ritterbush would say, or mere lifeless mechanical form, as A. W. Schlegel and Coleridge and many another Romantic would say.[13] So by a line of reasoning that starts with biological imagery we arrive at the conclusion that organic form can occur in visual art only by not occurring, at either terminus of a spectrum running from realistic representation to extreme abstraction.

If we believe that a poem grows in the mind like a plant (which is what Coleridge and the others did believe, or at least assert), and if we notice that the poem which emerges from the mind does not in fact look very much like a plant, and that furthermore (as we have been saying) the poem may or may not contain vegetable imagery, then as we ponder and expound our doctrine of growth and form, it may well be the perhaps more profound, but certainly less inspectable, part of the doctrine, the accent on the genetic, that we assert with the most energy. And thus it was in fact with Coleridge.

A few of his most striking and deliberate statements about organic form occur in his notes for a general lecture on Shakespeare, which editors rightly place in the context of yet other notes for lectures on Shakespeare's power as a poet, his imagination, his judgment so happily wedded with his genius. Here Coleridge executes a double step away from any possible implication that organic form consists in vegetable imagery. He moves the discussion into the fully human and dramatic arena of Shakespeare's plays and his *Venus and Adonis*. At the same time he is recurrently inclined to depart from the poems themselves and to search the organic depths of the mind of the great maker. This is Coleridge's well-known leaning. Wordsworth was content to illustrate the concept of 'imagination' in poems and passages of poems, especially in his own. Coleridge, as he himself explains in the *Biographia*, essayed the further radical task of tracing the poetic principle to its seat in the psychology of the poet.

Five properties of plant life (according to the clean exposition[14] by Meyer Abrams) enter into the analogy between plants and poems to be construed from Coleridge's several treatises, notes, letters, and conversations. (Coleridge both draws in part upon a German source, A. W. Schlegel, and in turn becomes archetypal for a moderate English tradition. Many passages in the first two volumes of Professor Wellek's *History of Modern Criticism* testify to the preoccupation of the Romantic Germans with organic form – and also to their extravagance.)[15] The five properties or principles of organic form, in the order arranged by Abrams, are these: 1. that of the WHOLE, the priority of the whole; without the whole the parts are nothing; 2. that of GROWTH, the manifestation of growth in the 'evolution and extension of the plant'; 3. that of ASSIMILATION; the plant converts diverse materials into its own substance; 4. that of INTERNALITY; the plant is the spontaneous source of its own energy; it is not shaped from without; 5. that of INTERDEPENDENCE, between parts and parts, and parts and whole; pull off a leaf and it dies. These somewhat overlapping or merging principles are all in effect equivalents of the single principle that we may call 'Organic Form'. The five, I believe, might be readily synthesized into fewer, or into the one; or they might be analyzed into a larger number. They have a close affinity for, or near identity with, a sixth, the favorite Coleridgean concept of the tension and reconciliation of manifold opposites which it is the peculiar power of artistic genius to accomplish.[16] All of these principles as expounded by Coleridge blend a measure of poetic structuralism, or objective doctrine concerning poetic form, with a measure of geneticism or psychological doctrine concerning the author's consciousness or unconsciousness. The second principle, or that of GROWTH, especially invites the genetic accent. As I have the authority of Professor Orsini on my side, I will not on this occasion take upon myself the full burden of the argument against what I have fallen into the habit of referring to as the 'intentional' or the 'genetic' fallacy. 'It is only to the finished product', says Professor Orsini, 'that we can apply the concept of organic unity.'[17] I assent emphatically. Let me add, however, one observation. There is at least one respect in which the physical organism, either growing plant or animal, is immeasurably surpassed by the human poetic consciousness. I mean, in its capacity for self-revision, rearrangement, mending. Plants renew leaves and flowers; animals moult in several ways; a lobster can lose a claw and regrow it; the human body heals cuts and regrows a finger nail. But there is no action of any physical organism that remotely approaches the power of the human mind to revise and recast *itself* – constantly to reaffirm or to cancel its own precedent action, in whole or in part. We confront here self-involution, a spiritual power. (The world soul, says Plotinus, looking to its consequent dreams up the physical universe; looking to its antecedent it reflects the ideas of the Nous.) As if a tree could move one of its own branches from the bottom to the top, or on looking itself over could change from an oak to a pine. What we call the 'finished product', the poem, is a moment of spiritual activity, hypostatized, remembered, recorded, repeated. The human psyche makes the poem out of itself, or offers a remembered action of itself as the poem. Thus it differs notably from the tree, which does not offer anything, but simply appears, as the necessary product of the process which is itself. The Romantic analogy between vegetable and poetic creation tended to assimilate the poetic to the vegetable by making the poetic as radically spontaneous as possible – that is, indeliberate, unconscious. Some theorists clearly affirmed this. Shakespeare created his Hamlet 'as a bird weaves its nest'.[18] A poet, urged Schiller, should *be* a plant.[19] An alternative which we have come close to noticing when we alluded to the eighteenth-century nature philosophers, was to draw plant life closer to human consciousness. According to one generous analogical view, plants *were* conscious. Coleridge, as Meyer Abrams has shrewdly pointed out, enjoyed the kind of classical sanity that compelled him to reject both solutions. Shakespeare's judgment was equal to his genius. He never wrote anything without conscious design.[20] On the other hand, 'the man would be a dreamer, who otherwise than poetically should speak of roses and lilies as *self-conscious* subjects.'[21] The inside history of literature as recorded in the testimonies of authors themselves is full of their awareness that the process by which they have arrived at the mental and verbal act presented as a poem has not necessarily, or even usually, been identical with that act as finally achieved. The moment presented as the poem is a contrived moment. This is so even on the supposition that the author achieves his sonnet in one perfect first draft. For he reviews it and accepts it and puts it out as a poem. No matter how spontaneous and lucky in one sense, in another sense it is also artificial. Few poets have, like the French inspirationalist Charles Peguy, looked on their first impulses as so literally inspired that the least revision of a first draft was an aesthetic sin. 'A line will take us hours maybe', says Yeats. 'Yet if it does not seem a moment's thought, Our stitching and unstitching has been naught.'[22] He knew that the hours of stitching and unstitching were a normal part of 'Adam's Curse'. And with a somewhat different emphasis: 'Verse, 'tis true,' argues Dryden, 'is not the effect of sudden thought; but that hinders not that sudden thought may be represented in verse.' 'A play is supposed to be the work of a poet.'[23]

And so we come to the third of three issues which I am trying to define – not whether the poem presents biological imagery; and not whether the process of its growth in the mind resembles the growth of a tree; but whether the poem itself, the hypostatized verbal and mental act, looks in any way like an animal or a vegetable. In Section 62 of the *Critique of Judgment* Kant observes, and few besides Professor Orsini seem to have noticed it,[24] that a work of human art differs from a natural organism in that the latter is self-organizing, that it can repair itself when damaged, and that it reproduces itself. I have already suggested that under its genetic aspect (as the creator's mind in act) the verbal work of art rivals and even surpasses the natural organism in the capacity of self-correction. We have seen that Coleridge, with his strong inclination to the genetic, claimed internality, or self-organization, as one of the characters of the poetic organism. But our theme now is the poem as presented or objectified act – as poetic object. Plato said a composition should have an organized sequence of parts, and that it ought to be *like* a living being, with foot, body, and head.[25] And Aristotle said that it ought to be a unity, like an organism.[26] But we might ask of Plato: What are the foot, body, and head of a poem? Or of Aristotle: What are the beginning, middle, and end of a squirrel or a tree? Or, Professor Orsini recites for us the rude question of an imagined objector: 'What corresponds to the stomach in a tragedy?' Such conceptions, he remarks wisely, are carrying the simile too far. He argues indeed that the simile (or metaphor) of physical organic life is not essential to the concept of aesthetic organic unity.[27] The aesthetic unity is generated by the Kantian *a priori* synthetic idea, the human reason's glorious power of non-empirical creative unifying vision.[28] The art work, says Professor Orsini, has indeed, and literally, an organic form, a synthetic unity in multiplicity. The merely physical organism enjoys this character only by metaphoric extension and hence in a less exact degree.[29] Thus he would reverse the usual direction of the metaphor. One will readily nowadays think of certain senses in which he may be right about the physical organism. Nowadays a batch of amoebas is chopped up and the parts are reassembled, more or less higgledy-piggledy, as I understand it, and a new set of amoebas emerges – 'synthetic'.[30] The human body, we read with queasy feelings, fights hard to reject the benevolently transplanted kidney or heart. Yet even this obtuse archaic organism (our body), struggling to carry out the Coleridgean rules, can be coerced for a certain time, even an extended time, into entertaining and being sustained by alien organs. And thus it succeeds in looking a little more than we might previously have thought possible like a machine with interchangeable parts.

The aesthetic organicist, therefore, in his dealing with poems, will no doubt do well to appeal but cautiously to that analogy with the all too ragged physical organism. He may well be content to confine his appeal to a very purified post-Kantian version of the aesthetic properties – the individuality and uniqueness of each aesthetic whole, the priority of the whole to the parts, the congruence and interdependence of parts with parts and of parts with the whole, the uniqueness and irreplaceability of parts and their non-existence prior to the aesthetic whole or outside it.[31] Surely these are ideas against which no literary critic is likely to rebel – none at least whose knowledge of critical history extends far enough backwards for him to appreciate the embarrassments for criticism created by the more extreme versions of legislation according to the classical literary kinds, or of evaluation according to the classical ornamental rhetoric, or of explanation according to economic, sociological, or other historical categories, or according to any theological, anthropological, or psychological archetypes. If we had never known any Romantic interest in life forms, if we had never heard of organic form, we should today be under the necessity of inventing it. We might well be dedicating this very volume to a struggle to invent and proclaim some doctrine of Romantic organicism. Given, however, the very well-established theory as we do know it, and given its several main articles of doctrine, such as we have been reciting, I will take this occasion to confess my opinion that both the metaphor and the literal idealist doctrine invite some not unreasonable questions.

If the leaf is detached from the tree, it dies. Still we may press it between the pages of a book and treasure it years later. It has that kind of superiority to certain parts of other and higher organisms, say an ear cut from a vanquished bull by a matador, a human finger cut off and preserved in formaldehyde. The German metaphysical humorist Jean Paul speaks in a typically enough romantic idiom when in his *Vorschule der Aesthetik* he alludes contemptuously to the traditional right of book-reviewers to pluck the feathers of the 'jewelled hummingbird'.[32] Jean Paul is confident too that 'the spirit of a work like the *Iliad* is manifest both in the whole and in the smallest syllable.'[33] 'Load every rift with ore', said Keats to Shelley. And even Coleridge, with a hedging glance at a Kantian distinction, uttered this well-known half-betrayal: A poem proposes to itself 'such delight from the *whole*, as is compatible with a distinct gratification from each component *part*'.[34] Few of us, I suppose, have had the experience of finding a hummingbird's tail feather. But I have found many blue-jay feathers, and pheasant feathers, and glossy black crow feathers – all of which I thought were beautiful and put in my hat or lapel or preserved for a while at home in a vase. No doubt there is an implicit sustaining context within which we admire such relics, and the same is true for certain fragmentary, sketchy, or partly abstracted representational forms of art – for instance, a single fleeting leg in a sculpture entitled 'Runner' by Leon Underwood.[35] We admire these in the context of an habitual consciousness of their relation to the rest of the visible surface of a bird or a human body. And these surface contexts too have beyond them an interior context – where interdependence or mutual need of parts is very great; it is peremptory or absolute. These we know about, and no doubt the fact has aesthetic significance. Still we do not need to see these things; we do not want to see them. The ancient haruspicators were bent on no aesthetic purpose. Elizabethan or Augustan lovers might hope to see themselves imaged through a window in a lady's bosom, or to look more cynically at the 'moving toy shop' of her heart, silks, ribbons, laces, gewgaws.[36] But when Humbert in Nabokov's *Lolita* longs in effect to eviscerate his nymphet, to kiss her insides, heart, liver, lungs, kidneys, we have already, some chapters before this, begun to enjoy a dawning comprehension that he is a madman.[37] I am trying to emphasize one dimension of gross differentiation between physical organism and poetic organism and thus to re-enforce the opinion I have already expressed that it is easy to push the analogy between them too far. We know there is not any part, detail, or aspect of a poem which we cannot at least try and wish to see in relation to all the other parts and the whole. The poem is all knowable; it is all knowledge, through and through. It is transnoetic – an act or a possible act of a self-reflexive consciousness. (In this respect, certain other kinds of art, stone statues, for instance, have a status different from that of either a poem or a person. A flaw at the center of the marble does not become known unless the statue is destroyed. Still the statue *is* solid and opaque and is conceived aesthetically in that way. Shelley's remark about the impurity of all the arts other than verbal poetry was more accurate than the more unifying idealisms.) And so we might at first think that the absolute idealist doctrines – no life in the part without the whole, no substitution of one part for another, and the like – if purged of too much contact with the unhappy biological metaphor and applied *literally* (as the idealist says) to poems, might hold up much better. I think this is in fact not true. I am not urging a paradox, but only one further confrontation with reality. I mean that in some respects the poem as organic unity will come off rather worse without the crutch, or the distraction, of the physical comparison.

The head of many a marble statue has survived truncation – and has been admired for centuries in a garden or a museum. Necessity in the relation of parts to parts and of parts to whole differs very widely with different arts and with different parts. Go into a movie by Antonioni or Fellini just in time to see the last few flickers, and you will likely experience a nearly maximum loss of meaning and aesthetic quality in a part deprived of its whole. Arrive at the beginning of the movie, or read the first chapter of almost any novel, and you will likely have the opposite experience. If the high aesthetic doctrine of the whole were true, we would never sit out even a very good movie or a very good play, never finish reading a novel or a poem. We recognize and enjoy the trenchancy or the delicacy of Augustan couplet wit before we finish the first page of any one of the major poems in that mode. We recognize and reject the sentimental inflation, the witless couplets, of Erasmus Darwin's *Loves of the Plants* on reading a very few lines. The reader unsophisticated by aesthetic theory has a constant and not always unhappy tendency to escape the tyranny of title pages, chapter, act, and scene headings, even the tag of the *dramatis personae*. The wider stretches of poetry are often, like life, a kind of spread-out and general, or atmospheric, or virtual, context for local episodes of the most intense aesthetic quality. Many couplets by Alexander Pope are better poems in their own right than Ezra Pound's miniature image *In A Station of the Metro*. Matthew Arnold's 'touchstones' do not offer a viable method of criticism, but his conception of them is far from absurd. A very different spirit, in a far distant era, the Roman neoplatonist Plotinus spoke words of wisdom against the Stoic notion that symmetry is one of the necessary conditions of the beautiful. Think, he says, what that doctrine leads us to. 'Only a compound can be beautiful, never anything devoid of parts . . . Yet beauty in an aggregate demands beauty in details; it cannot be constructed out of ugliness; its law must run throughout.'[38]

We are told by the Kantian aesthetician, solemnly, that the interdependence of parts in the organically unified poem is so close that to remove any part is to 'damage' the whole (to damage it badly, we suppose), and that no part is replaceable by any other conceivable part.[39] But how can we ever be sure about either of these propositions? Many poets continue to revise their poems assiduously, to remove parts, to add others, to replace others – even to the last gasp of their deathbed editions. Alexander Pope's intended order of satiric portraits in his moral Epistle *Of the Characters of Women* remains conjectural today. It apparently remained unclear to him (and to Warburton) what the ideal order was. Some of the portraits apparently have been lost.[40] Editors, compositors, critics, theatrical producers, and actors make many inspired changes in works – either on purpose or accidentally. (Richard Burton has turned his back on the audience, lurked in the shadows, and mumbled the soliloquies of *Hamlet* – in effect actor against the play.) Even a Dunce, Lewis Theobald, introduced an emendation into Shakespeare – which for two centuries has been gratefully accepted. Startling examples of this kind, from the annals of textual editing, from critical speculation, from innocent appreciation directed to a corrupt text, have been assembled in a recent article by Professor James Thorpe, who was pushing the case for the author against the printer's devil and his associates.[41] We need have no quarrel with this cause. We are aiming here, not at genetic or textual problems, but at a confrontation with certain perhaps embarrassing aesthetic dubieties. Five-act plays and epic narratives are often lumpy – in ways that producers of plays can cope with but which reading aestheticians may have to blink. The authors are masters of episodes and scenes. Think of the partition of the kingdom by Hotspur, Glendower, and Mortimer in the first part of *King Henry IV*, of Justice Shallow's nostalgia for the good old days at Clement's Inn in the second part of the same chronicle play. Think of the closing books of the *Odyssey*. Think of the Doloneia or night-time slaughter of the scout in Book X of the *Iliad* – perhaps genetically intrusive, as scholiasts for centuries have suspected – in any event a hypertrophic development, but a cherished one.[42] A professor of aesthetics, Catherine Lord, has argued that too close a degree of organic unity necessarily defeats the episodic and multifarious nature of such extended literary works.[43] An advance speculatist in Renaissance studies, Harry Berger, notices the 'conspicuous irrelevance' of descriptive detail, the 'perverse insistence on the digressive elements', with which Spenser roughens and gives character to the otherwise too smooth and logical lines of allegory in Book II of *The Faerie Queene*.[44] It is my own heretical belief that a good chess problem, viewed according to the idealist organistic norm, has a more fully determined and hence more perfect structure than even a sonnet by Shakespeare. A hallmark of linguistic expression, as linguists are now telling us, is a certain surplus of information. The *Hopscotch* novel, with a hundred expendable chapters to be inserted at intervals as the reader wills,[45] and other sorts of open-ended or multiple-choice fictions, are now a well-established feature of the literary scene. Such terms as 'indeterminacy', 'irrelevance', and 'nonstructure' and studies devoted to *Strains of Discord* begin to appeal to the critics. We need not be partisans of all these kinds of innovation in order to carry on our dialogue with Coleridge and Professor Orsini.

In saying all this, we are, of course, subscribing without reserve to the Kantian proposition recently cited by Professor Hirsch,[46] that we must confront and interpret the aesthetic object in our best frame of mind. Whatever is ideal (if it is not in fact clearly chimerical) is what in fact the work is and says. This extends (in the absence of other kinds of factual evidence) to textual arguments about intentions. What makes clearly better sense always has the superior claim. Hamlet, as F. W. Bateson has sensibly affirmed, yearned not that 'this too, too sullied flesh', but that 'this too, too *solid* flesh would melt, thaw, and resolve itself into a dew.'[47] The Wife of Bath, as Talbot Donaldson, manifesting both textual erudition and concern for poetic meaning, has been the only editor to conclude, speaks of human organs of generation having been created not by a 'perfectly wise wight', but by a 'perfectly wise wright' – *a conditore sapientissimo*, as the Wife's source St Jerome had put it. And we are convinced, not because the key word happens to occur in three 'bad' manuscripts among the total of fifty-two, but because this is the only reading that makes good sense.[48]

The direction in which my argument has been pointing must be clear. Examples of less than complete organicity such as I have cited could be multiplied indefinitely and indefinitely varied. But I seem to hear the neo-Kantian idealist voices murmuring: 'Enough of this.

You are talking of imperfections. The organicist doctrine applies only to aesthetic perfection, and perfection in this world is hard to come by.'[49] And I answer: Just so. But I think of passages in Plato where he seems to be voting in favor of poetry, or at least in favor of poetic inspiration – if only that inspiration will produce something like the beautiful wisdom of philosophy. The poets as we know them, Homer and the tragedians, are a mad gang of corrupters. They are outside the pale. I myself have been speaking of English poetry as we know it – of Shakespeare, for instance, or Pope. A doctrine of organicity, if it means an exceedingly subtle, intimate, manifold (and hence dramatic and imaginative) 'interinanimation' of parts in a poem, must surely be one of the modern critic's most carefully defended doctrines. Yet if he faces the facts, he will at the same time find the organic structure of the poem, perhaps paradoxically, a notably loose, stretchable and adjustable kind of organic form. A 'loose' conception of poetic organicism is, in short, what I am arguing for. The time has perhaps arrived in the dialectic of literary theory when we gain little by repeating the organistic formulas. We can perhaps gain more now by trying to test and extend the more precise schemata which are at our disposal for describing the organization of poems. Some advances, along with perhaps some merely ingenious exercises, in modes of grammatical exegesis are being shown these days by critics of the 'structural' inclination – and notably by those of the orientation toward Paris. Brilliant exercises, for instance, have been broadcast by Roman Jakobson and a few associates. Sonnets by Baudelaire, Sidney, Shakespeare, a song by Blake along with its illumination, verses by two painters, the *douanier* Rousseau and Paul Klee, are subjected to extremes of analysis under the rigorous structural technique, and they yield no doubt some subliminal secrets of 'the grammar of poetry and the poetry of grammar'. A similarly progressive, a finely tempered and well-assured idiom of analysis is demonstrated in a recent book (1968) by an American scholar, Barbara Herrnstein Smith. The title will suggest something of the special insight: *Poetic Closure: A Study of How Poems End.*[50] Stephen Booth's *Essay on Shakespeare's Sonnets* is an even more recent adventure (1969) which, despite what to my mind is an over-emphasis on the reader's 'experience' of a sonnet (in addition to the sonnet itself), takes a more or less rewarding interest in such logical and grammatical commonplaces as 'Unity and Division, Likeness and Difference', and in various multiple occurrences of pattern – 'Formal, Logical, and Syntactical', 'Rhetorical', 'Phonetic', and 'Lexical'.[51]

Taken together, these two books exhibit an ingenious array of structural commentary upon English lyric poems – Elizabethan, Metaphysical and Cavalier, Romantic, and post symbolist. Explication of poems is, to my mind, one of the termini of literary criticism. It should rarely, if ever, be reported or re-explicated in anybody else's essay. It suits my purpose of the moment very well, however, to take some notice of the theoretical or speculative idiom employed by Mrs Smith.

> A poem or a piece of music concludes. We tend to speak of conclusions when a sequence of events has a relatively high degree of structure (p. 2).

> If, on the other hand, there have been no surprises or disappointments, if all our expectations have been gratified, then the poem has been as predictable – and as interesting – as someone's reciting the alphabet. Art inhabits the country between chaos and cliché (p. 14).

> We may think of *integrity* as, in one sense, the property of a system of which the parts are more obviously related to each other than to anything outside that system (pp. 23–4).

> Closure . . . may be regarded as a modification of structure that makes *stasis*, or the absence of further continuation, the most probable succeeding event. Closure allows the reader to be satisfied by the failure of continuation or, put another way, it creates in the reader the expectation of nothing (p. 34).

> This does not mean that our experience of the work ceases abruptly at the last word. On the contrary, at that point we should be able to re-experience the entire work, not now as a succession of events, but as an integral design. The point may be clarified if we consider that we cannot speak of the 'end' of a painting or a piece of sculpture (p. 36).

How shall we describe or locate this unpretentiously lucid and persuasive idiom of generalization about poems? It is of course Mrs Smith's own idiom – an achievement which has helped to earn her book at least two prizes and a number of encomiastic reviews. At the same time, we can identify it with some exactitude, I believe, as a judicious blend of *Gestalt* psychology, which Mrs Smith acknowledges, and of Aristotelian common sense, which strangely she is silent about. We might ascribe this to a prudent strategy by which the author is taking care not to look in the least old-fashioned. I incline rather to ascribe it to absentmindedness – and the fairly close resemblance in some of the phrasing simply to the principle that very good ideas, classically simple, essential, and true ideas, are likely to crop up spontaneously in any age – even in the midst of crowding rival fantasies and fads. Aristotle, as we have noted, does, in his account of the literary object, make a momentary appeal to the analogy of biological organism.[52] But for him it is indeed a momentary analogy and no more. A wholesome lesson that we can derive from this juncture in critical history – both from Aristotle and from Mrs Smith and others of her like who are these days raising their voices – is that neither the organicism of the extreme biological analogy nor that of the *a priori* or transcendental absolute assertion is likely to encourage superior readings of poetry, but rather that homelier and humbler sort of organicism, in the middle, which I have been trying to describe – empirical, tentative, analytic, psychological, grammatical, lexicographic. This, I believe, was in effect the kind of organicism which was the preoccupation of the American critics who were chronologically 'new' a third of a century ago but who were, or are indeed, both as old and as new as mankind's literate ambition to make as much sense as possible of the perennially experienced, muddled shape of things.

Notes

1 See, for example, Earl R. Wasserman, *The Subtler Language*, Baltimore: The Johns Hopkins University Press, 1959, ch. 4.

2 Erasmus Darwin, *The Botanic Garden* (1789–91); 4th edition, London, 1799, Part II, *The Loves of the Plants*, 9 . . . 20, 31–8. Cf. Philip C.

Ritterbush, *Overtures to Biology*, New Haven: Yale University Press, 1964, pp. 162–5.

3 *Overtures*, p. 156.

4 *Biographia Literaria*, ed. J. Shawcross, London: Oxford University Press, 1939, II, p. 257 ('On poesy or art').

5 *Biographia Literaria*, II, pp. 253–4, 258.

6 Ritterbush, *The Art of Organic Forms*, City of Washington: Smithsonian Institution Press, 1968, pp. 19–20.

7 *The Art of Organic Forms*, p. 23. René Wellek, *A History of Modern Criticism*, III, 1963, p. 140, speculates that Ruskin's liking for organistic theory influenced his distaste for Dutch painting and classical landscapes.

8 Maurice Rheims, *The Age of Art Nouveau*, trans. Patrick Evans, London: Thames & Hudson, 1966, p. 14; cf. p. 95; Plate 114.

9 No doubt to be found somewhere in Geffroy's eight vols of *La Vie artistique*, 1892–1903. I quote from Martin Battersby, *The World of Art Nouveau*, New York: Funk & Wagnalls, 1968, p. 145. Cf. Rheims, op. cit., p. 212.

10 *The Art of Organic Forms*, pp. 83–4.

11 Ibid., p. 86.

12 Josef Albers, *Despite Straight Lines*, New Haven: Yale University Press, 1961, p. 10.

13 'These results require the use of ruler and drafting pen and establish unmodulated line as a legitimate artistic means. In this way they oppose a belief that the handmade is better than the machine-made, or that mechanical construction is anti-graphic or unable to arouse emotion.' Albers, op. cit., p. 16.

14 Meyer Abrams, *The Mirror and the Lamp*, London: Oxford University Press, 1953, pp. 170–6.

15 See, for instance, II, pp. 48, 358: A. W. Schlegel, in his Berlin Lectures, said that Euripides was the 'putrefaction of Greek tragic form'.

16 René Wellek, 'Coleridge's philosophy and criticism', in *The English Romantic Poets, A Review of Research*, ed. T. M. Raysor, New York: Modern Language Association of America, 1950, pp. 109, 113, argues the close connection for Coleridge between the organic principle and that of polarity of opposites; he thinks too sharp a distinction between the two is drawn by Gordon McKenzie, *Organic Unity in Coleridge*, Berkeley, 1929.

17 G. N. G. Orsini, 'The organic concepts in aesthetics', *Comparative Literature*, XXI (winter, 1969), p. 5.

18 Wellek, *History*, III, p. 166 (Emerson); Cf. II, p. 290 (Kleist); II, p. 217 (A. W. Schlegel).

19 Abrams, pp. 168–74; 202–8.

20 Abrams, pp. 364–5. It was still possible for Walter Pater to read the

general emphasis of Coleridge's *dicta* on organicism in the opposite way. He thought that Coleridge made the artist 'almost a mechanical agent'; poetry 'like some blindly organic process of assimilation' ('Coleridge', *Appreciations* [London, 1898], p. 80). And indeed see 'On poesy or art', *Biographia*, II, p. 258: 'There is in genius itself an unconscious activity; nay, that is the genius in the man of genius.'

21 Abrams, p. 173 and note to pp. 364–5.

22 'Adam's Curse', *In the Seven Woods*, 1904.

23 John Dryden, Essays, ed. W. P. Ker, I, pp. 102, 114: *Essay of Dramatic Poesy* and *Defence of the Essay*.

24 G. N. G. Orsini, *Coleridge and German Idealism*, Carbondale, Illinois: Southern Illinois University Press, 1969, pp. 160–2.

25 Plato, *Phaedrus*, 265.

26 Aristotle, *Poetics*, VIII, 4; XXIII, 1.

27 Orsini, pp. 4–5, 27. Cf. Wellek, *History*, I, pp. 9, 18, 26 (the romantic emphasis upon biology); IV, pp. 70–1 (Brunetière adopting biological evolutionary concepts too literally); Meyer Abram, 'Archetypal analogues in the language of criticism', *University of Toronto Quarterly*, XVIII (1949), pp. 313–27; Graham Hough, *An Essay on Criticism*, London, 1966, ch. 22, 'Organic form: a metaphor'.

28 The reference is to Kant's *Logik*, ch. 1, para. 3 (Orsini, pp. 2–3, 4, 17). This kind of unity embraces also mental activities other than the aesthetic – e.g. philosophical, political, scientific, technological (Orsini, p. 26). Though the idealist aesthetician must guard against the sin of 'intellectualism' in defining the unifying motifs of aesthetic works (Orsini, pp. 9–10), he is ready to extend the purified concept of 'organic' unity into areas of the highest abstractionism. This, I should say, involves Plotinian and Crocean problems about how to distinguish art from the whole remaining horizon of being and of human knowing.

29 Orsini, p. 27.

30 *New York Times*, 13 November 1970, p. 1.

31 Orsini, pp. 17, 10–11.

32 *Vorschule* (1804–13), Second Preface, para. 11, in *Sämtliche Werke*, ed. Eduard Berend, XI, Part I (Weimar, 1935), p. 11. I am indebted to the translation by Margaret Cardwell Hale, forthcoming from the Wayne State University Press.

33 Ch. 86, p. 311. The quotation is from Mrs Hale's close paraphrase of Jean Paul's fragmentary chapter.

34 *Biographia*, II, p. 10 (ch. 14). He was hedging less carefully when he said that poetry permits a 'pleasure from the whole consistent with a consciousness of pleasure from the component parts', and that it communicates 'from each part the greatest sum of pleasure compatible with the largest sum of pleasure of the whole' ('Definition of poetry', *Shakespearian Criticism*, ed. T. M. Raysor, I, p. 148).

35 R. H. Wilenski, *The Meaning of Modern Sculpture*, London: Faber & Faber, 1932, Plate 18 (facing p. 133) and p. 160.
36 Murray Krieger, *A Window to Criticism*, Princeton University Press, 1964, Part II; *The Rape of the Lock*, I, 100; and Pope's *Guardian*, 106.
37 Vladmir Nabokov, *Lolita*, New York: G. P. Putnam's Sons, 1955, p. 167.
38 *Enneads*, I, vi, 1.
39 'In the unified object, everything that is necessary is there, and nothing that is not necessary is there . . . if a certain yellow patch were not in a painting, its entire character would be altered, and so would a play if a particular scene were not in it, in the place where it is . . . in a good melody, or in painting, or poem, one could not change a part without damaging (not merely changing) the whole' (John Hospers, 'Problems of aesthetics', *Encyclopaedia of Philosophy* [New York, 1967], I, pp. 43–44, quoted by Orsini, pp. 1–5).
40 Frank Brady, 'The History and Structure of Pope's *To a Lady*', *Studies in English Literature 1500–1900*, IX (summer, 1969), pp. 439–62: '*To a Lady* is no immutable "organic" whole.'
41 'The aesthetics of textual criticism', *PMLA*, LXXX (1965), pp. 465–82. Cf. Fredson Bowers, 'Textual criticism', *The Aims and Methods of Scholarship in Modern Languages and Literatures*, ed. James Thorpe, New York: Modern Language Association of America, 1963, p. 24.
42 Cedric Whitman, *Homer and the Heroic Tradition*, Cambridge, Massachusetts: Harvard University Press, 1958, pp. 283–4, 353.
43 Catherine Lord, 'Organic unity reconsidered', *JAAC*, XXII (spring, 1964), pp. 263–8, N. 17, p. 268, quotes Hans Eichner: 'Shakespeare is the very last dramatist whose plays one would normally describe as integrated wholes' ('The meaning of "Good" in aesthetic judgments', *The British Journal of Aesthetics* (October, 1963), p. 316, n. 3).
44 Harry Berger, *The Allegorical Temper, Vision and Reality in Book 2 of Spenser's Faerie Queene*, New Haven: Yale University Press, 1953, chs 5–7; esp. pp. 122–3, 128.
45 Julio Cortázar, *Hopscotch*, trans. from the Spanish (*Razuela*) by Gregory Rabana, New York: Pantheon Books, 1966.
46 Kant, *Critique of Judgment*, 21; E. D. Hirsch, 'Literary evaluation as knowledge', *Contemporary Literature*, IX (summer, 1968), p. 328.
47 'Modern bibliography and the literary artifact', in *English Studies Today*, ed. G. A. Bonnard, Bern: Francke Verlag, 1961, pp. 67–70.
48 E. Talbot Donaldson, *Speaking of Chaucer*, New York: W. W. Norton & Co., 1970, pp. 115, 119–21, 125–8.
49 Orsini, p. 3.
50 Barbara Herrnstein Smith, *Poetic Closure, A Study of How Poems End*, University of Chicago Press, 1968.
51 Stephen Booth, *An Essay on Shakespeare's Sonnets*, New Haven: Yale University Press, 1969.
52 Aristotle, *Poetics*, XXIII, 1.

G. S. Rousseau

A selective bibliography of works on organic form

Few bibliographies are exhaustive, and this one does not even claim to be. It has been included partly because no similar tool exists – at least no published bibliography that the editor and authors can discover; but also indebted to the insistence of recent writers on the subject (e.g. G. N. Giordano Orsini, Launcelot L. Whyte, Ludwig von Bertalanffy) that such a checklist, however inexhaustive, be compiled. Surely its present form of names and titles chronologically arranged is somewhat dead, but the alternative – a critical bibliography with a few descriptive sentences after each title, or even a single comment – would have swelled this work to Brobdingnagian proportions. The editor very much hopes that such a critical bibliography will be prepared in the not too distant future.

While the advantages of chronological arrangement are perfectly clear, why start in 1823 with an article by Henri Milne-Edwards in the *Archives générales de Médecine*? Because thinking on the subject of organic form up to the early 1820s has been admirably surveyed by Dr Ritterbush in his essay, and because we wished to effect as smooth as possible a transition from the essays to the bibliography. But also because we had to begin somewhere and unless we surveyed this vast field in its entire chronological landscape, from the ancient Greeks to 1970, an arbitrary starting point had to be selected. Finally, the early 1820s were chosen to demonstrate, however perfunctorily, the development of thought about organic form throughout the nineteenth century. To have plunged in at the end of the century when there were, admittedly, increasingly more items being written in the subject, would have left the reader without a context and, additionally, created a sizable lacuna from the time of Goethe's death in 1832. Since Goethe is so very central to any reconsideration of organic form, his lifetime offered reasonable boundaries of chronology by which to begin the arduous process of selection and elimination.

Finally, by what criteria were selections made and why are there gaps between certain dates? (e.g., 1824, 1831, 1836, etc.) The field was surveyed as exhaustively as possible for English, French, German, and American writers. Unfortunately it was not possible – again for reasons of size – to include the Italians, Spaniards, or East Europeans. After this, writings were included in the printed list only if they were believed to say something new on the subject or to have influenced contemporaries or later writers. If the twentieth century seems more copious in its listings than the nineteenth, or again if the nineteen fifties and sixties more than the thirties and forties, this is not due to an overweighting in the present or to galloping scribosis – for which arguments could be advanced – but to an increasing interest by men of many walks of life. While organic form as an early nineteenth-century topic was primarily confined to biologists and creative thinkers, today (and more generally in recent times) it finds students among social scientists, systems analysts, urban planners, and legislators, in addition to scientific and humanistic communities where one would naturally expect it to be discussed. Recent listings have therefore been more exhaustively noted to present the range of this body of writings.

Surely it would be superfluous to comment here on the changes in thinking about organic form, especially as they occurred decade by decade, or from thinker to thinker. Dr Ritterbush has explored some of these changes, and any remarks made here would be too cursory to be meaningful. Scrutiny of the bibliography, however, shows that thinking about organic form mirrors the composite best thinking of an age, whether it be the aftermath of the French Revolution, the prelude to both World Wars, or the explosive (i.e. communication explosion) sixties. For in some ways the recent history of this idea is also the recent history of human culture: so large and all-embracing is the notion of organic form that it transcends any restricted application to literary criticism and theoretical biology. But most salient is the way the idea has become 'international' in this century. No longer confining itself to thinkers of a particular geography or school of thought, the problems raised by the idea of organic form are vital problems for citizens of the whole world. And this fact is rendered abundantly clear by the following bibliography.

Note: Abbreviations of journals are the standard ones listed in the *Annual Bibliography of the Publications of the Modern Language Association of America.*

1823 MILNE-EDWARDS, Henri, 'Mémoire sur la structure élémentaire des principaux tissus organiques des animaux', *Archives générales de médecine*, 1st year, III, 165–84.

1824 DUTROCHET, Henri Joachim, *Recherches anatomiques et physiologiques sur la structure intime des animaux et des végétaux, et sur leur mobilité*, Paris.

1831 BRAUN, Alexander, 'Vergleichende Untersuchung über die Ordnung der Schuppen an den Tannenzapfen als Einleitung zur Untersuchung der Blattstellung', *Nova Acta Physico-Medica Academiae Caesareae Leopoldino-Carolinae*, XVII, 195–402.

1836 SCHNECKENBERGER, T. Ch., *Ueuber die Symmetrie der Pflanzen: Eine Inaugural-Dissertation*, Tübingen.

1838 EHRENBERG, Christian Gottfried, *Die Infusionsthierchen als vollkommene Organismen: Ein Blick in dos tiefere organische Leben der Natur*, Leipzig.

1840 DAVY, Sir Humphry, 'Parallels between art and science' [1807] in *The Collected Works of Sir Humphrey Davy, Bart . . .*, ed. John Davy, VIII, London, 306–9.

1841 DUJARDIN, Felix, *Histoire naturelle des zoophytes: Infusoires, comprenant la physiologie et la classification de ces animaux, et la manière de les étudier à l'aide du microscope*, Paris.

1847 SCHWANN, Theodore, *Microscopical Researches into the Accordance in the Structure and Growth of Animals and Plants*, trans. Henry Smith, London.

1856 JONES, Owen, *The Grammar of Ornament*, London.

1866 HAECKEL, Ernst, *Generelle Morphologie der Organismen: Allgemeine Grundzüge der organische Formen Wissenschaft, mechanisch begründet durch die von Charles Darwin reformirte Descendenz-Theorie*, 2 vols, Berlin.

1870 BEALE, Lionel S., *Protoplasm: or, Life, Matter, and Mind*, 2nd edition, London.

1870 MACALISTER, Alexander, 'On the law of symmetry as exemplified in animal forms', *Journal of the Royal Dublin Society*, V, 326–38.

1873 REINKE, Johannes, *Morphologische Abhandlungen*, Leipzig.

1876 BÜTSCHLI, Otto, 'Studien über die ersten Entwicklungsvorgänge der Eizelle, die Zelltheilung und die Conjugation der Infusorien', *Abhandlungen herausegegeben von der senckengergischen naturforschenden Gesellschaft*, X, 213–464.

1876 HAECKEL, Ernst, *The History of Creation: Or the Development of the Earth and its Inhabitants by the Action of Natural Causes*, trans. E. Ray Lankester, 2 vols, New York.

1880 MONTGOMERY, Edmund, 'The Unity of the Organic Individual', *Mind*, V, xix, 318–36; V, xx, 465–89.

1881 CANDOLLE, Casimir de, *Considérations sur l'étude de la phyllotaxie*, Geneva.

1884 HALDANE, J. S., 'Life and Mechanism', *Mind*, IX, 27–47.

1886 BERTHOLD, Gottfried, *Studien über Protoplasmamechanik*, Leipzig.

1890 BÜTSCHLI, Otto, *Ueber den Bau der Bacterien und verwandter Organismen*, Leipzig.

1892 BÜTSCHLI, Otto, *Untersuchungen über mikroskopische Schaüme und das Protoplasma: Versuche und Beobachtungen zur Lösung der Frage nach den physikalischen Bedingungen der Lebenserscheinungen*, Leipzig.

1894 DRIESCH, Hans, *Analytische Theorie der organischen Entwicklung*, Leipzig.

1897 ANDREWS, Gwendolen Foulke, 'The living substance: as such and as organism', *Journal of Morphology*, supplement to XII.

1898 BÜTSCHLI, Otto, *Untersuchungen über Strukturen insbesondere über Strukturen nichtzelliger Erzengnisse des Organismus und über ihre Beziehungen zu Strukturen, welche ausserhalb des Organismus entstehen*, Leipzig.

1898 WILSON, Edmund B., *The Cell in Development and Inheritance*, New York.

1899, 1904 HAECKEL, Ernst, *Kunst-Formen der Natur*, Leipzig and Vienna.

1901 REINKE, Johannes, *Einleitung in die theorefische Biologie*, Berlin.

1905 HAECKEL, Ernst, *The Wonders of Life: A Popular Study of Biological Philosophy*, trans. Joseph MacCabe, New York.

1908 DRIESCH, Hans, *The Science and Philosophy of the Organism*, London.

1908 MOEBIUS, Carl, *Asthetik der Tierwelt*, Jena.

1908 PETTIGREW, J. Bell, *Design in Nature: Illustrated by Spiral and Other Arrangements in the Inorganic and Organic Kingdoms As Exemplified in Matter, Force, Life, Growth, Rhythm, etc., Especially in Crystals, Plants, and Animals*, 3 vols, London.

1910 COKER, F. W., 'Organismic theories of the State. Nineteenth century interpretations of the State as organism or as person', in *Columbia University Studies in History, Economics and Public Law*, XXXVIII, ii.

1912 COLMAN, Samuel, *Nature's Harmonic Unity: A Treatise on Its Relation to Proportional Form*, ed. C. Arthur Coan, New York.

1914 COOK, Theodore Andrea, *The Curves of Life: Being an Account of Spiral Formations and Their Application to Growth in Nature, to Science and to Art with Special Reference to the Manuscripts of Leonardo da Vinci*, New York.

1914 JOHNSTONE, James, *The Philosophy of Biology*, Cambridge, England.

1914 LEDUC, Stéphane Armand-Nicolas, *The Mechanism of Life*, trans. W. Deane Butcher, New York.

1915 HAMBURGER, Margarete, *Das Form-problem in der neuren deutschen Ästhetik und Kunsttheorie*, Heidelberg.

1916 HACK, Ray Kenneth, 'The Doctrine of Literary Forms', *Harvard Studies in Classical Philology*, XXVII, 35–6.

1916 LOEB, Jacques, *The Organism as a Whole, from a Physicochemical Viewpoint*, New York.

1916 RUSSELL, E. S., *Form and Function: A Contribution to the History of Animal Morphology*, London.

1916 SANDAY, William, *Form and Content in the Christian tradition: a friendly discussion between W. Sanday . . . and N. P. Williams*, London.

1917 HAECKEL, Ernst, *Kristallseelen: Studien über das anorganische Leben*, Leipzig.

1917 HENDERSON, L. J., *The Order of Nature*, Harvard.

1917 THOMPSON, D. W., *On Growth and Form*, Cambridge, England.

1919 REINKE, Johannes, *Die schaffende Natur. Mit Bezugnahme auf Schopenhauer und Bergson*, Leipzig.

1919 WHITEHEAD, Alfred North, *The Concept of Nature: the Tarner Lectures delivered in Trinity College, November 1919*, Cambridge, England.

1920 HAMBRIDGE, Jay, *Dynamic Symmetry: The Greek Vase*, New Haven, Connecticut.

1920 JAEGER, F. M., *Lectures on the Principle of Symmetry and its Application in all Natural Sciences*, London.

1923 LEWIS, Frederick T., 'A note on symmetry as a factor in the evolution of plants and animals', *American Naturalist*, LVII, 5–41.

1923 LEWIS, Frederick T., 'The typical shape of polyhedral cells in vegetable parenchyma and the restoration of that shape following cell division', *Proceedings of the American Academy of Arts and Sciences*, LVIII, 537–52.

1924 DÜRKEN, Bernard, *Allgemeine Abstammungslehre*, Berlin.

1925 BARNES, H. E., 'Representative biological theories of society', *Sociological Review*, XVII, 182–94; 294–300.

1925 BEWS, John William, *Plant Forms and their Evolution in South Africa*, London.

1926 DRIESCH, Hans, *Philosophie des Organischen*, Leipzig.

1926 NEEDHAM, Joseph, 'S. T. Coleridge as a philosophical biologist', *Science Progress*, XX, lxxx, 692–702.

1928 BERTALANFFY, Ludwig von, *Kritische Theorie der Formbildung*, Berlin.

1928 CROW, W. B., 'Symmetry in organisms', *American Naturalist*, LXII, 207–27.

1928 KER, William Paton, *Form and Style in Poetry: Lectures and Notes, by W. P. Ker*, ed. R. W. Chambers, London.

1928 NEEDHAM, Joseph, 'Organicism in biology', *Journal of Philosophical Studies*, III, ix, 29–40.

1929 BLOSSFELDT, Karl, *Art forms in Nature: Examples from the Plant World Photographed Direct from Nature* [1928], New York.

1929 HALDANE, J. S., *The Sciences and Philosophy*, London.

1929 NEEDHAM, Joseph, 'Protein metabolism and organic evolution', *Science Progress*, XXIII, xcii, 633–48.

1930 FRIEDMANN, H., *Welt der Formen*, Munich.

1930 RUYER, R., *Esquisse d'une philosophie de la structure*, Paris.

1930, 1931 WOODGER, Joseph Henry, 'The concept of "organism" and the relation between embryology and genetics', *Quarterly Review of Biology*, V, 1–22, 438–62; VI, 178–207.

1931 HALDANE, J. S., *The Philosophical Basis of Biology*, New York.

1931 OZENFANT, Amedée, *Foundations of Modern Art*, London.

1932 BERTALANFFY, Ludwig von, *Theoretische Biologie*, I, Berlin.

1932 HUXLEY, J. S., *Problems of Relative Growth*, London.

1932 READ, Herbert, *Form in Modern Poetry*, London.

1933 BERTALANFFY, Ludwig von, *Modern Theories of Development*, trans. J. H. Woodger, London.

1933 EVANS, Joan, *Nature in Design: A Study of Naturalism in Decoration Art from the Bronze Age to the Renaissance*, Oxford.

1934 CARNAP, R., *The Unity of Science*, London.

1934 COLD SPRING HARBOUR, 'Aspects of Growth' in *Symposia on Experimental Biology*, II, Cold Spring Harbour, U.S.A.

1934 PROCHNOW, Oskar, *Formenkunst der Natur*, Berlin.

1935 CALKINS, G. N., *The Smallest Living Things*, London.

1935 HALDANE, J. S., *The Philosophy of a Biologist*, Oxford.

1935 WATKIN, Edward Ingram, *A Philosophy of Form*, New York.

1936 BARR, Alfred H., Jr, *Cubism and Abstract Art*, New York.

1936 DÜRKEN, Bernard, *Entwicklungsbiologie und Ganzheit*, Leipzig and Berlin.

1936 GERRITY, Brother Benignus, *The Relations between the Theory of Matter and Form and the Theory of Knowledge in the Philosophy of Saint Thomas Aquinas*, Washington.

1936 LEVY, Julien, *Surrealism*, New York.

1936 LUCAS, F. L., *The Decline and Fall of the Romantic Ideal*, New York.

1936 NEEDHAM, Joseph, *Order and Life*, Cambridge, England and New Haven, Connecticut.

1936 SYZ, H., 'The concept of the organism-as-a-whole and its application to clinical situations', *Human Biology*, VIII, iv, 489–507.

1938 BERTALANFFY, Ludwig von, 'Inquiries on growth laws II: a quantitative theory of organic growth', *Human Biology*, X, 181–213.

1938 CARNAP, R., 'Logical foundations of the unity of science' in *International Encyclopedia of Unified Science*, I, Chicago, 42–62.

1938 FREY-WYSSLING, Albert, *Submikroskopische Morphologie des Protoplasmas und seiner Derivate*, Berlin.

1938 MCDOUGALL, William, *The Riddle of Life: a Survey of Theories*, London.

1938 WEAVER, Bennett, 'Wordsworth: forms and images', *SP*, XXXV, 433–45.

1939 GILBERT, Katharine Everett and KUHN, Helmut, *A History of Esthetics*, New York.

1939 KONCZEWSKA, Hélène, *L'Unité de la matière et le problème des transmutations*, Paris.

1939 MCKENZIE, Gordon, 'Organic unity in Coleridge' in *University of California Publications in English*, VII, i.

1939 SINNOTT, Edmund W., 'The cell and the problem of organization', *Science*, LXXXIX, 41–6.

1940 BROOKS, Cleanth, 'The poem as organism', *English Institute Annual*, 20–41.

1940 PICKEN, L. E. R., 'The Fine Structure of Biological Systems', *Biological Reviews of the Cambridge Philosophical Society*, XV, 133–67.

1940 SCHMID, Gunther, *Goethe und die Naturwissenschaften: Eine Bibliographie*, Halle.

1941 GODE-von AESCH, Alexander, 'Natural science in German Romanticism', *Columbia University Germanic Studies*, n.s., no. 11.

1941 LOVEJOY, A. O., 'The meaning of Romanticism for the historian of ideas', *JHI*, II, iii, 257–78.

1941 NIERENDORF, Karl, *Paul Klee: Paintings, Watercolors 1913 to 1939*, New York.

1942 GUGGENHEIM, Marguerite, ed., *Art of This Century . . .* with Introductory essays by André Breton, 'Genesis and Perspective of Surrealism'; Hans Arp, 'Abstract Art, Concrete Art'; and Piet Mondrian, 'Abstract Art', New York.

1942 NEEDHAM, Joseph, *Biochemistry and Morphogenesis*, Cambridge, England.

1943 AGAR, W. E., *A Contribution to the Theory of the Living Organism*, Melbourne.

1944 JANIS, Sidney, *Abstract and Surrealist Art in America*, New York.

1944 JOHNSON, M., *Art and Scientific Thought*, London.

1944 KEPES, Gyorgy, *Language of Vision*, Chicago.

1945 BRETON, André, *Le Surréalisme et la peinture suivi de genèse et perspective artistiques du surréalisme et de fragments inédits*, New York.

1945 CLARK, W. E. Le Gros and MEDAWAR, P. B., eds, *Essays on Growth and Form Presented to D'Arcy Wentworth Thompson*, Oxford.

1945 LILLIE, Ralph Stayner, *General Biology and Philosophy of the Organism*, Chicago.

1945 RUSSELL, Edward Stuart, *The Directiveness of Organic Activities*, Cambridge, England.

1945 SCHRÖDINGER, E., *What is Life?*, Cambridge, England.

1945 ZINK, Sidney, 'The poetic organism', *Journal of Philosophy*, XLII, xvi, 421–34.

1946 ARBER, Agnes, 'Goethe's botany', *Chronica Botanica*, X, 63–124.

1946 CARLES, Jules, *Unité et vie; esquisse d'une biophilosophie*, Paris.

1946 GHYKA, Matila, *The Geometry of Art and Life*, New York.

1947 KEETON, Morris T., 'Edmund Montgomery – pioneer of organicism', *JHI*, VIII, iii, 309–41.

1947 LIENAU, C. C., 'Quantitative aspects of organization', *Human Biology*, XIX, 163–216.

1948, 1949, 1952, 1953, 1955 BAKER, John R., 'The cell theory: a restatement, history, and Critique', *Quarterly Journal of Microscopical Science*, LXXXIX, 103–25; XC, 87–108, 331; XCIII, 157–89; XCIV, 407–40; XCVI, 449–81.

1948 ERNST, Max, 'Beyond painting and other writings by the artist and his friends' in *The Documents of Modern Art*, ed. Robert Motherwell, New York.

1948 HOLMES, Samuel Jackson, *Organic Form and Related Biological Problems*, Berkeley.

1948 SMITH, R. Jack, 'Intention in an organic theory of poetry', *Sewanee Review*, LVI, 625–33.

1949 BERTALANFFY, Ludwig von, *Das biologische Weltbild*, I, Bern.

1949 BERTALANFFY, Ludwig von, *Vom Molekül zur Organismenwelt*, Potsdam.

1949 BERTALANFFY, Ludwig von, 'Problems of organic growth', *Nature*, CLXIII, 156–8.

1949 WHYTE, Lancelot L., *The Unitary Principle in Physics and Biology*, London.

1950 ARBER, Agnes, *The Natural Philosophy of Plant Form*, London.

1950, 1951 BASS, Robert E., 'Abstracts', *Physical Rev.*, LXXIX, 201; LXXXI, 295.

1950 BATE, Walter Jackson, 'Coleridge on the function of art' in *Perspectives of Criticism*, ed. Harry Levin, Cambridge, England.

1950 BERTALANFFY, Ludwig von, 'The theory of open systems in physics and biology', *Science*, CXI, iii, 23–9.

1950 GRAVE, S. A., 'Aristotelian philosophy and functional design', *Australian Journal of Philosophy*, XXVIII, 29–42.

1950 GREENE, Theodore M., 'The scope of aesthetics' (Part II of 'Is a general theory of arts of any practical value in the study of literature?'), *JAAC*, VIII, 221–8.

1950 JESSUP, BERTRAM, 'Aesthetic size', *JAAC*, IX, 31–8.

1950 NEWTON, Eric, *The Meaning of Beauty*, New York.

1950 SINNOTT, Edmund W., *Cell and Psyche; The Biology of Purpose*, Chapel Hill.

1950 SVOBODA, K., 'Content, subject and material of a work of literature', *JAAC*, IX, 39–45.

1950 WHYTE, Lancelot L., *The Next Development in Man*, New York.

1950 WIGOD, J. D., 'The unity of "Ode on a Grecian Urn" ', *Personalist*, XXXI, 149–56.

1951 BASS, Robert E., 'Unity of nature', in 'General system theory: a new approach to unity of science', *Human Biology*, XXIII, iv, 323–7.

1951 BENZIGER, James, 'Organic unity: Leibnitz to Coleridge', *PMLA*, LXVI, 24–48.

1951 BERTALANFFY, Ludwig von, 'Conclusion', in 'General system theory: a new approach to unity of science', 336–45.

1951 BERTALANFFY, Ludwig von, 'Goethe's concept of nature', *Main Currents in Modern Thought*, VIII, 78–83.

1951 BERTALANFFY, Ludwig von, 'Problems of general system theory', in 'General system theory: a new approach to unity of science', 302–12.

1951 BERTALANFFY, Ludwig von, 'Towards a physical theory of organic teleology: feedback and dynamics', in 'General system theory: a new approach to unity of science', op. cit., 346–61.

1951 DÉRIBÉRÉ, Maurice, *Images étranges de la nature*, Paris.

1951 DESTOUCHES, Paulette Février, *La structure des théories physiques*, Paris.

1951 GIBSON, James J., 'What is a form?', *Psychological Review*, LVIII, 403–12.

1951 HEMPEL, C. G., 'General system theory and the unity of science', in 'General system theory: a new approach to unity of science', 313–22.

1951 HILL, Archibald A., 'Towards a literary analysis', *University of Virginia Stud.*, IV, 147–65.

1951 JONAS, Hans, 'Comment on general systems theory', in 'General systems theory: a new approach to unity of science', 328–35.

1951, 1952 KRISTELLER, Paul Oskar, 'The modern system of the arts: a study in the history of aesthetics', *JHI*, XII, 496–527; XIII, 17–46.

1951 MOHOLY-NAGY, Sibyl, 'Idea and pure form', *Arts and Architecture*, LXVIII, iii, 24–5, 46.

1951 NAGEL, E., 'Mechanistic explanation and organismic biology', *Philosophy and Phenomenological Research*, XI, 327–38.

1951 SCHIBUBERT-SOLDERN, R., *Philosophie des Lebendigen*, Graz.

1951 WHYTE, Lancelot L., ed., *Aspects of Form; a Symposium on Form in Nature and Art*, Bloomington, Indiana.

1952 BÜNNING, E., 'Morphogenesis in plants', *Survey of Biological Progress*, II, 105–40.

1952 CANGUILHEM, Georges, *La Connaissance de la vie*, Paris.

1952 NAGEL, E., 'Wholes, sums, and organic unities', *Philosophical Studies*, III, ii, 17–32.

1952 PORTMANN, Adolf, *Animal Forms and Patterns; A Study of the Appearance of Animals*, London.

1952 *Proceedings, Symposium on the Biochemical and Structural Basis of Morphogenesis*, Utrecht.

1952 READ, Herbert, 'Farewell to formalism', *Art News*, LIII, 36–9.

1952 UREY, H. C., *The Planets, Their Origin and Development*, New Haven, Connecticut.

1952 WEYL, Hermann, *Symmetry*, Princeton, New Jersey.
1952 WOODGER, Joseph Henry, *Biology and Language: An Introduction to the Methodology of the Biological Sciences Including Medicine*, Cambridge, England.
1953 ABRAMS, M. H., *The Mirror and the Lamp: Romantic Theory and the Critical Tradition*, New York.
1953 CASSIRER, Ernst, *The Philosophy of Symbolic Forms*, trans. Ralph Manheim, New Haven, Connecticut.
1953 LANGER, Susanne K., *Feeling and Form*, New York.
1953 STEINBERG, Leo, 'The eye is a part of the mind', *Partisan Review*, XX, 194–212.
1954 ADAMS, Richard P., 'Emerson and the organic metaphor', *PMLA*, LXIX, 117–30.
1954 FISHMAN, Solomon, 'Sir Herbert Read: poetics vs. criticism', *JAAC*, XIII, 156–62.
1954 HAMM, Victor M., 'The problem of form in nature and the arts', *JAAC*, XIII, 175–84.
1954 MUNRO, Thomas, 'The morphology of art as a branch of aesthetics', *JAAC*, XII, 438–49.
1954 READ, Herbert, 'Art and the evolution of consciousness', *JAAC*, XIII, 143–55.
1954 SMITH, Cyril Stanley, 'The shape of things', *Scientific American*, CXC, 58–64.
1954 TILLYARD, E. M., 'Shakespeare's historical cycle: organicism or compilation?', *SP*, LI, 34–9.
1954 UNGERER, Emil, *Die Wissenschaft vom Leben; eine Geschichte der Biologie*, Freiburg.
1954 VAN FLEET, D. S., 'Cell and tissue differentiation in relation to growth (plants)', *Soc. Devel. and Growth Symposium*, XI, 111–29.
1954 WHYTE, Lancelot L., *Accent on Form: an Anticipation of the Science of Tomorrow*, New York.
1955 ADAM, Robert M., 'Literature and psychology: a question of significant form', *Literature and Psychology*, V, 67–72.
1955 FOGLE, Richard Harter, 'Organic form in American criticism: 1840–1870', in *The Development of American Literary Criticism*, ed. Floyd Stovall, Chapel Hill, 75–111.
1955 HINTIKKA, Kaarlo Jaakko Juhani, *Two Papers on Symbolic Logic: Form and Content in Quantification Theory, and Reductions in the Theory of Types*, Helsinki.
1955 JOHNSON, Wendell Stacy, 'Some functions of poetic form', *JAAC*, XIII, 496–506.
1955 KOHLSCHMIDT, Werner, *Form und Innerlichkeit; Beiträge zur Geschichte und Wirkung der deutschen Klassik und Romantik*, Bern.
1955 KRIEGER, Murray, 'Benedetto Croce and the recent poetics of organicism', *Comparative Literature*, VII, 252–8.

1955 MUNRO, Thomas, 'Form and value in the arts: a functional approach', *JAAC*, XIII, 316–41.
1955 SINNOTT, Edmund W., *The Biology of the Spirit*, New York.
1955 VIVAS, Eliseo, *Creation and Discovery: Essays in Criticism and Aesthetics*, New York.
1955 WEISS, Paul, 'Beauty and the beast: life and the rule of order', *The Scientific Monthly*, LXXXI, 286–99.
1956 AMES, Van Meter, 'What is Form?', *JAAC*, XV, 85–93.
1956 BACON, Roseline, *Odilon Redon*, 2 vols, Geneva.
1956 BEARDSLEY, Monroe C., 'The concept of economy in art', *JAAC*, XIV, 370–5.
1956 BROWN, Theodore M., 'Greenaigh, Paine, Emerson and the organic aesthetic', *JAAC*, XIV, 304–17.
1956 FEININGER, Andreas, *The Anatomy of Nature: How Function Shapes the Form and Design of Animate and Inanimate Structures Throughout the Universe*, New York.
1956 FISHMAN, Solomon, 'Meaning and structure in poetry', *JAAC*, XIV, 453–61.
1956 FUSSINER, Howard, 'Organic integration in Cézanne's painting', *College Art Journal*, XV, 302–12.
1956 KAHN, Sholom J., 'Towards an organic criticism', *JAAC*, XV, 58–73.
1956 MANUEL, Frank E., 'From equality to organicism', *JHI*, XVII, i, 54–69.
1956 STRACHE, Wolf, *Forms and Patterns in Nature*, New York.
1956 WELLEK, René, 'Coleridge's philosophy and criticism', in *The English Romantic Poets*, ed. T. M. Raysor, New York, 110–37.
1956 WHYTE, Lancelot L., 'Some thoughts on the design of nature and their implication for education', *Arts and Architecture*, LXXIII, 16–17.
1956 WOODGER, Joseph Henry, *Physics, Psychology, and Medicine: A Methodological Essay*, Cambridge, England.
1957 BURKE, Kenneth, *The Philosophy of Literary Form*, New York.
1957 GERARD, R. W., 'Units and concepts of biology', *Science*, CXXV, 429–33.
1957 GÉRARD, Albert, 'On the logic of Romanticism', *Essays in Criticism*, VII, iii, 262–73.
1957 LA DRIÈRE, James C., '"Form" and "Expression": meaning and structure in language and in art', in *Atti del III° Congresso Internazionale di Estetica*, 497–502.
1957 MOULYN, Adrian C., *Structure, Function and Purpose; An Inquiry into the Concepts and Methods of Biology from the Viewpoint of Time*, New York.
1957 PAULING, Linus C., ed., *Molecular Structure and Biological Specificity*, Washington D.C.
1957 SCHNIER, Jacques, 'The function and origin of form', *JAAC*, XVI, 66–75.
1957 SIOHAN, Robert, 'Une micro-organisme sonore', in *Atti del III° Congresso Internazionale di Estetica*, 647–50.

1957 WEIDLÉ, Wladimir, 'Biology of art: initial formulation and primary orientation', trans. E. P. Halperin, *Diogenes*, no. 17, 1–15.
1958 BAUR, John I. H., *Nature in Abstraction*, New York.
1958 HOMMA Hans, *Das Formproblem in der Biologie*, Vienna.
1958 KELLY, George, 'Poe's theory of unity', *PQ*, XXXVII, 34–44.
1958 RUYER, Raymond, *La Genèse des formes vivantes*, Paris.
1958 SCHMIDT, Georg and SCHENK, Robert, *Kunst und Naturform*, Basle.
1958 WELLEK, René, 'Concepts of form and structure in twentieth-century criticism', *Neophilologus*, XLII, 2–11.
1959 BAHM, Archie J., 'Matter and spirit: implications of the organicist view', *Philosophy and Phenomenological Research*, XX, i, 103–8.
1959 CROOK, E., *The Structure and Function of Subcellular Components*. Cambridge, England.
1959 HUGHES, Arthur, *A History of Cytology*, London.
1959 LECKY, Eleazer, 'Ideas of "Order" in modern literary criticism', *Literature and Psychology*, IX, 36–9.
1959 SCHULZ, M. F., 'Oneness and multeity in Coleridge's poems', *Tulane Studies in English*, IX, 53–60.
1959 WIGGLESWORTH, V. B., *The Control of Growth and Form: A Study of the Epidermal Cell in an Insect*, Ithaca, New York.
1960 BALTRUŠAITIS, Jurgis, *Réveils et prodiges: le gothique fantastique*, Paris.
1960 FLORKIN, Maurice, *Naissance et déviation de la théorie cellulaire dans l'oeuvre de Theodore Schwann*, Paris.
1960 KEPES, Gyorgy, ed., *The Visual Arts Today*, Middletown, Connecticut.
1960 KUHNS, Richard F., 'Art structures', *JAAC*, XIX, 91–8.
1960 READ, Herbert, *The Forms of Things Unknown*, London.
1960 SEWELL, Elizabeth, *The Orphic Voice: Poetry and Natural History*, New Haven, Connecticut.
1960 SINNOTT, Edmund W., *Plant Morphogenesis*, New York.
1960 VATTIMO, Gianni, 'Opera d'arte e organismo in Aristotle', *Revista di Estetica*, V, 358–82.
1960 WEISS, Paul, 'Organic form: scientific and aesthetic aspects', *Daedalus*, LXXXIX, 177–90.
1961 BLUM, H. F., 'On the origin and evolution of living machines', *American Scientist*, XLIX, 474–501.
1961 BOYD, J. D., JOHNSON, F. R. and LEVER, J. D., eds, *Electron Microscopy in Anatomy*, Baltimore.
1961 FISCHER, Trudy, IRONS, Ian and FISCHER, Roland, 'Patterns in art and science', *Studies in Art Education*, II, 85–100.
1961 GOODWIN, T. W. and LINDBERG, O., eds, *Biological Structure and Function*, New York.
1961 SCHAPER, Eva, 'Significant form', *British Journal of Aesthetics*, I, 33–43.
1961 WADDINGTON, Conrad H., *The Nature of Life*, London.
1962 BLINDERMAN, Charles S., 'T. H. Huxley's theory of aesthetics: unity in diversity', *JAAC*, XXI, 49–55.

1962 FARÉ, Michel, *La Nature morte en France: son histoire et son évolution du XVIIe à XXe siècle*, Geneva.

1962 FOGLE, Richard Harter, 'Coleridge on organic unity: life'; 'Coleridge on organic unity: beauty'; 'Coleridge on organic unity: poetry', in 'The idea of Coleridge's criticism', *Perspectives in Criticism*, IX, 18–69.

1962 PURVES, A. C., 'Formal structure in "Kubla Khan" ', *SIR*, I, 187–91.

1962 RUSSELL, Edward Stuart, *The Diversity of Animals: An Evolutionary Study*, Leiden.

1962 SIEGEL, Curt, *Structure and Form in Modern Architecture*, trans. Thomas E. Burton, New York.

1962 SIMON, Herbert, 'The architecture of complexity', *Proceedings, American Philosophical Society*, CVI, 467–82.

1962 STAHL, W. R., 'Similarity and dimensional methods in biology', *Science*, CXXXVII, 205–12.

1962 WADDINGTON, Conrad H., *New Patterns in Genetics and Development*, New York.

1963 ALLEN, John M., *The Nature of Biological Diversity*, New York.

1963 BENDMANN, A., 'Die organismische Auffassung Bertalanffy's', *Deutsche Zeitschrift für Philosophie*, XI, 216–22.

1963 BOULEAU, Charles, *The Painter's Secret Geometry: A Study of Composition in Art*, trans. Jonathan Griffin, New York.

1963 CANGUILHEM, Georges, 'The role of analogies and models in biological discovery', in *Scientific Change: Historical Studies in the Intellectual, Social and Technical Conditions for Scientific Discovery and Technical Invention, from Antiquity to the Present*, ed. Alistair C. Crombie, New York, 507–20.

1963 CUNNINGHAM, J. V., 'The problem of form', *Shenandoah*, XIV, ii, 3–6.

1963 HENCHMAN, Michael J., 'Guts, positrons and the origin of life: inherent asymmetry in the physical world', *Journal of the Leeds University Chemical Society*, V, 51–60.

1963 KAHLER, Eric, 'The forms of form', *Centennial Review*, VII, 131–43.

1963 MAYER, Edmund, *Introduction to Dynamic Morphology*, New York.

1963 NISSÈN, Claus, 'Über Botanikmalerei', *Atlantis*, XXXV, 349–68.

1963 PERGER, Anton, *Analogien in unserem Weltbild; Entwurf einer allgemeiner Formenlehre*, Meisenheim am Glan.

1963 PURCELL, Edward, 'Parts and wholes in physics', in *Parts and Wholes*, ed. Daniel S. Lerner, New York.

1963 RHODIN, Johannes A. G., *An Atlas of Ultrastructure*, Philadelphia.

1963 RICHARDS, I. A., 'How does a poem know when it is finished?', in *Parts and Wholes*, ed. Daniel S. Lerner, New York.

1963 SALVADORI, Mario George, *Structure in Architecture*, Englewood Cliffs.

1963 SINNOTT, Edmund W., *The Problem of Organic Form*, New Haven, Connecticut.

1964 ARBER, Agnes, *The Mind and the Eye: A Study of the Biologist's Standpoint*, London.

1964 GARDNER, Martin, *The Ambidextrous Universe*, New York.

1964 GUTMAN, Herbert, 'Structure and function', *Genetic Psychology Monographs*, LXX, 3–56.

1964 KOESTLER, Arthur, *The Act of Creation*, New York.

1964 LORD, Catherine, 'Organic unity reconsidered', *JAAC*, XXII, 263–8.

1964 MILLER, Craig W., 'Coleridge's concept of nature', *JHI*, XXV, 77–96.

1964 ORSINI, G. N. G., 'Coleridge and Schlegel reconsidered', *Comparative Literature*, XVI, 99–118.

1964 POLANYI, Michael, *Personal Knowledge: Towards a Post-Critical Philosophy*, New York.

1964 PORTER, Keith R. and BONNEVILLE, Mary A., *An Introduction to the Fine Structure of Cells and Tissues*, Philadelphia.

1964 RITTERBUSH, Philip C., *Overtures to Biology: The Speculations of Eighteenth-Century Naturalists*, New Haven, Connecticut.

1964 SMITH, Cyril Stanley, 'Structure, substructure, superstructure', *Reviews of Modern Physics*, XXXVI, 524–32.

1965 BARRY, David G., *Art in Science*, Albany, New York.

1965 DENNIS, Jane M. and WENNEKER, Lee B., 'Ornamentation and the organic architecture of Frank Lloyd Wright', *Art Journal*, XXV, 2–14.

1965 HARRIS, Errol E., *The Foundations of Metaphysics in Science*, New York.

1965 HUTCHINGS, P., 'Organic unity revindicated?', *JAAC*, XXIII, 323–8.

1965 HUTCHINSON, G. Evelyn, 'The naturalist as art critic', in *The Ecological Theater and the Evolutionary Play*, New Haven, Connecticut, 95–108.

1965 KEPES, Gyorgy, *Structure in Art and in Science*, New York.

1965 LECHEVALIER, Hubert A. and SOLOTOROVSKY, Morris, *Three Centuries of Microbiology*, New York.

1965 MENDELSOHN, Everett, 'Physical models and physiological concepts: explanation in nineteenth-century biology', *British Journal for the History of Science*, II, 201–19.

1965 MILLER, James G., 'Living systems: basic concepts, structure and process, cross-level hypotheses', *Behavioral Science*, X, 193–237; 337–441.

1965 NEEDHAM, Arthur Edwin, *The Uniqueness of Biological Materials*, Oxford.

1965 WHYTE, Lancelot L., 'Atomism, structure and form', in *Structure in Art and in Science*, ed. G. Kepes, New York.

1966 ALEXANDER, Christopher, 'A city is not a tree', *Design*, CCVI, 46–55.

1966 ALEXANDER, Christopher, *Notes on the Synthesis of Form*, Cambridge, Massachusetts.

1966 ALTMAN, Joseph, *Organic Foundations of Animal Behavior*, New York.

1966 BENNETT, Thomas Peter, *Modern Topics in Biochemistry; Structure and Function of Biological Molecules*, New York.

1966 BURKE, John G., *Origins of the Science of Crystals*, Berkeley and Los Angeles.

1966 BURKE, Kenneth, 'Formalist criticism: its principles and limits', *Texas Quarterly*, IX, 242–68.

1966 CAMPBELL, Peter Nelson, *The Structure and Function of Animal Cell Components; an Introductory Text*, New York.

1966 CAPITAN, William H., 'On unity in poems', *Monist*, L, 188–203.

1966 GOMBRICH, E. H., 'Meditations on a hobby horse, or the roots of artistic form', in *Aspects of Form*, Bloomington, Indiana and London, 209–28.

1966 GREGORY, F. G., 'Form in plants', in *Aspects of Form*, Bloomington, Indiana and London, 57–76.

1966 HARTMAN, Geoffrey, 'Beyond formalism', *MLN*, LXXXI, 542–56.

1966 HUMPHREYS-OWEN, S. P. F., 'Physical principles underlying inorganic form', in *Aspects of Form*, Bloomington, Indiana and London, 8–22.

1966 MALOF, Joseph, 'Meter as organic form', *MLQ*, XXVII, 3–17.

1966 NEEDHAM, Joseph, 'Biochemical aspects of form and growth', in *Aspects of Form*, Bloomington, Indiana and London, 77–90.

1966 OSBORNE, H., 'Wittgenstein on aesthetics', *British Journal of Aesthetics*, VI, 385–90.

1966 RIESER, Max, 'Problems of artistic form: the concept of form', *JAAC*, XXV, 17–26.

1966 ROOSEN-RUNGE, Peter, 'Toward a theory of parts and wholes: an algebraic approach', *General Systems Yearbook*, XI, 13–18.

1966 SINNOTT, Edmund W., *The Bridge of Life; From Matter to Spirit*, New York.

1966 SLATTERLY, Sister Mary Francis, 'Formal specification', *JAAC*, XXV, 83–8.

1966 WADDINGTON, Conrad H., 'The character of biological form', in *Aspects of Form*, Bloomington, Indiana and London, 43–56.

1966 WADDINGTON, Conrad H., *Principles of Development and Differentiation*, New York.

1966 WOLSTENHOLME, G. E. W. and O'CONNOR, Maeve, eds, *Principles of Biomolecular Organization*, CIBA Foundation Symposium, London.

1967 ALLEN, John M., ed., *Molecular Organization and Biological Function*, New York.

1967 BERNAL, J. D., 'Symmetry of the genesis of form', *Journal of Molecular Biology*, XXIV, 379–90.

1967 ENGSTRÖM, Arne, *Biological Ultrastructure*, New York.

1967 FARBER, Eduard, 'Concentric circles and chemistry', *Smithsonian Journal of History*, II, 31–42.

1967 FISCHER, Ernst, 'Chaos and Form', *Mosaic: A Journal for the Comparative Study of Literature and Ideas*, I, 132–40.

1967 GLACKEN, Clarence J., *Traces on the Rhodian Shore: Nature and Culture in*

Western Thought from Ancient Times to the End of the Eighteenth Century, Berkeley and Los Angeles.

1967 GOTSHALK, D. W., 'Form and expression in Kant's aesthetics', *British Journal of Aesthetics*, VII, 250–60.

1967 LANGER, Susanne K., Mind: *An Essay on Human Feeling*, I, Baltimore, Maryland.

1967 LE GRAND, Yves, *Form and Space Vision*, Bloomington, Indiana.

1967 LENNBERG, Eric H., *Biological Foundations of Language*, New York.

1967 STEMPEL, Daniel, 'Coleridge and organic form: The English tradition', *SIR*, VI, 89–97.

1967 THOMAS, Russell, 'Unity in the Arts', *Journal of General Education*, XIX, 35–47.

1967 WATHEN-DUNN, Weiant, ed., *Models for the Perception of Speech and Visual Form; Proceedings of a Symposium*, Cambridge, Massachusetts.

1968 BAHM, Archie J., 'The Aesthetics of Organicism', *JAAC*, XXV, 449–60.

1968 BECKNER, Morton, *The Biological Way of Thought*, Berkeley and Los Angeles.

1968 BERTALANFFY, Ludwig von, *General System Theory: Foundations, Developments, Applications*, New York.

1968 BERTALANFFY, Ludwig von, *Organismic Psychology and Systems Theory*, Worcester, Massachusetts.

1968 DRONAMRAJU, K. R., ed., *Haldane and Modern Biology*, Baltimore, Maryland.

1968 KOESTLER, Arthur, *The Ghost in the Machine*, New York.

1968 MIKÁROVSKÝ, Jan, 'The notion of the wholeness in the theory of art', *ES*, V, 173–83.

1968 RITTERBUSH, Philip C., *The Art of Organic Forms*, Washington D.C.

1968 RITTERBUSH, Philip C., 'The biological muse', *Natural History*, LXXVII, 6, 26–31, 84.

1968 SMITH, Christopher Upham Murray, *Molecular Biology; A Structural Approach*, Cambridge, Massachusetts.

1968 WADDINGTON, Conrad H., *Towards a Theoretical Biology*, Edinburgh.

1969 BLANDINO, Giovanni S. T., *Theories on the Nature of Life*, New York.

1969 ORSINI, G. N. G., 'The organic concepts in aesthetics', CL, XXI, 1–30.

1969 RIESER, Max, 'Problems of artistic form: the concept of art', *JAAC*, XXVII, 261–9.

1969 TILLMAN, Frank A. and CAHN, Steven M., eds, *Philosophy of Art and Aesthetics from Plato to Wittgenstein*, New York.

1969 WHYTE, Lancelot L., WILSON, Albert G. and WILSON, Donna, eds, *Hierarchical Structures*, New York.

1970 NEEDHAM, Joseph, *The Chemistry of Life: Eight Lectures on the History of Biochemistry*, Cambridge, England.

1970 PHILLIPS, D. C., 'Organicism in the late nineteenth and early twentieth centuries', *JHI*, XXI, iii, 413–32.

1970 ROSENBLUETH, Arturo, *Mind and Brain: A Philosophy of Science*, Cambridge, Massachusetts.

1970 SNELDERS, H. A. M., 'Romanticism and Naturphilosophie and the inorganic natural sciences 1797–1840: an introductory survey', SIR, IX, 193–215.

1970 WADDINGTON, Conrad H., *Behind Appearance: A Study of the Relations Between Painting and the Natural Sciences in This Century*, Cambridge, Massachusetts.

Index to bibliography

General index